MICROPOROUS AND MESOPOROUS MATERIALS

Edited by **Reza Sabet Dariani**

Microporous and Mesoporous Materials
http://dx.doi.org/10.5772/61556
Edited by Reza Sabet Dariani

Contributors

Oscar Anunziata, Marcos B. Gómez Costa, Juliana M. Juárez, Mohindar Singh Seehra, Vishal Narang, Afrodita Zendelska, Mirjana Golomeova, Adebola Akinjokun, T.V. Ojumu, Beata Michalkiewicz, Joanna Sreńscek-Nazzal, Urszula Narkiewicz, Antoni W. Morawski, Rafał Wróbel, Rafael Huirache-Acuña, Barbara Pawelec, Rufino Nava, Eric M. Rivera-Muñoz, Trino Zepeda, Juan Carlos Moreno Piraján

Published by InTech

Janeza Trdine 9, 51000 Rijeka, Croatia

Publishing Process Manager Romina Rovan
Technical Editor SPi Global
Cover InTech Design team

First published August, 2016
Printed in Croatia

Additional hard copies can be obtained from orders@intechopen.com

Microporous and Mesoporous Materials, Edited by Reza Sabet Dariani
p. cm.
Print ISBN 978-953-51-2582-2
Online ISBN 978-953-51-2583-9

For EU product safety concerns:
IN TECH d.o.o., Prolaz Marije Krucifikse Kozulić 3, 51000 Rijeka, Croatia,
info@intechopen.com or visit our website at intechopen.com.

Meet the editor

Dr. Reza Sabet Dariani completed his PhD in Physics at the age of 32 from the University of New South Wales and Postdoctoral position from the Queen's University. He is now Professor of Physics at the Alzahra University and Editor-in-Chief of the Alzahra University Journal of Applied Physics. He received the Third World Academy of Science (TWAS) grants for Spare Parts for Scientific Equipment, Trieste, Italy in 2002, and was Winner of the SVC best poster, The Society of Vacuum Coaters, Washington DC in 2006. He has published more than 97 papers in reputed journals and 113 conference papers, and is also a reviewer of several international journals.

Contents

Preface

Recently, a lot of attention has been given to microporous and mesoporous materials. The microporous and mesoporous materials are classified depending on their pore size up to 2 nm, and pore size 2 to 50 nm, respectively. The aim of this book has been to explore the variety of phenomena associated with the major forms of the material, while laying the foundation for a clear and detailed working and understanding of the materials. We tried to present new types of advanced materials, which are currently a hot topic, and provide readers with a selective review of important improvements in the field.

Accordingly, this book "Microporous and Mesoporous Materials" focuses on important advances presented in the following chapters: "Biomass, Abundant Resources for Synthesis of Mesoporous Silica Material", "Microporous and Mesoporous Materials in Decontamination of Water Process", "Overview of Phosphorous Effect in Molybdenum-Based Hydrotreating Catalysts Supported on Ordered Mesoporous Siliceous Materials", "Synthesis and Characterization of CMK Porous Carbons Modified with Metals Applied to Hydrogen Uptake and Storage", "Mesoporous Carbons for Energy-efficient Water Splitting to Produce Pure Hydrogen at Room Temperature", "The Increase of the Micoporosity and CO2 Adsorption Capacity of the Commercial Activated Carbon CWZ-22 by KOH Treatment", and "Application of Some Natural Porous Raw Materials for Removal of Lead and Zinc from Aqueous Solutions".

I believe that every chapter in this book presents the progress in the subject and describes the latest advances in microporous and mesoporous materials.

I am thankful to Ms. Romina Rovan, Publishing Process Manager, for her generous support at critical stages of this project. I also wish to acknowledge with gratitude the support and hospitality of InTech, Croatia.

Dr. Reza Sabet Dariani
Professor of Physics
Alzahra University
Tehran
Iran

Chapter 1

The Increase of the Micoporosity and CO_2 Adsorption Capacity of the Commercial Activated Carbon CWZ-22 by KOH Treatment

Joanna Sreńscek-Nazzal, Urszula Narkiewicz, Antoni W. Morawski, Rafał J. Wróbel and Beata Michalkiewicz

Additional information is available at the end of the chapter

http://dx.doi.org/10.5772/63672

Abstract

The chemical modification of CWZ-22—commercial activated carbon (AC) with KOH- to enhance CO_2 adsorption was examined. The effect of different impregnation ratios KOH:CWZ-22 from 1 to 4 was studied. The ACs were characterized by CO_2 and N_2 sorption, Fourier transform infrared (FTIR), SEM, and XRD methods.

The impregnation of CWZ-22 with KOH highly effectively increased the porosity, specific surface area, and pore volume of ACs. The specific surface area of KOH:CWZ-22 = 4 increased to 1299 m^2/g as compared with pristine, which is equal to 856 m^2/g. The total pore volume raised from 0.51 to 0.77 cm^3/g.

The chemical modification of CWZ-22 increased the CO_2 adsorption capacity up to 38%. The Sips model described very good CO_2 adsorption on KOH:CWZ-22 = 4 and pristine sample. The surface of both ACs was homogenous. The values of isosteric heats of adsorption indicated on physisorption.

Keywords: carbon dioxide, activated carbon, CWZ-22, KOH, CO_2 adsorption

1. Introduction

The emission of CO_2 originating mainly from industry has become a worldwide problem responsible for the global warming. The combustion of fossil fuel in power plants remains the main point source for CO_2 emission to the atmosphere. Reduction in the CO_2 concentration in the atmosphere is currently a hot topic [1, 2]. Among technologies proposed for reduction in CO_2 emissions, adsorption is considered as a very promising process for CO_2 capture. Alternatively, solid-based adsorbents have drawn substantial attention for CO_2 capture.

Nowadays, many types of porous materials have been used in CO_2 adsorption, such as zeolites [3], metal–organic frameworks (MOFs) [4], porous silica [5], and activated carbons (ACs) [2, 6–9]. Among these adsorbents, activated carbon (AC) has drawn great attention recently because of its high adsorption capacity, low cost, availability, large surface area, an easy-to-design pore structure, hydrophobicity (insensitiveness to moisture), and low energy requirements for regeneration [10, 11]. The adsorption performance of activated carbons depends on the selection of carbon sources and activation conditions. It has been reported that the activated carbons prepared with commercial carpet [12], eucalyptus sawdust [13], yeast [14], palm shells [10], peanut shell [8], pitch [9], and molasses [15] had high adsorption capacity for CO_2.

Activated carbon can be mainly prepared by physical and chemical activation methods or by combination of both types of methods. Usually, physical activation is carried out using carbon dioxide, steam, air, or their mixture. Chemical activation involves agents such as zinc chloride [16], acids [17], and bases [10, 17]. KOH is one of the most widely used chemicals for activating the carbonaceous materials in preparation of activated carbon [1, 2, 8, 10]. Activated carbons obtained by chemical activation often possess a high specific surface area and well-developed micropores, which make them attractive materials for CO_2 adsorption. Particularly, KOH activation has been applied in the preparation of activated carbons because it can produce lots of micropores favorable for CO_2 adsorption. Therefore, the textural properties of the activated carbons depend on a type of carbon sources and the required amount of KOH for the preparation of efficient activated carbon to adsorb CO_2.

Our motivation was to increase the porosity and CO_2 adsorption capacity of commercial activated carbon CWZ-22. The increase of the CO_2 adsorption on commercial activated carbon modified using KOH has not been yet described.

2. Experimental method

2.1. Materials and sample preparation

A commercial activated carbon CWZ-22 (Gryfskand Sp. z o.o. Hajnówka, Poland) was used as the raw material in this work. The CWZ-22 samples were mixed with the saturated KOH solution. The mass ratio of KOH:CWZ-22 was varied from 1 to 4. The soaking time was 3 h. The mixtures were dried at 200°C. The impregnated sample was activated at temperature of 800°C for 1 h under nitrogen flow. Then, the samples were washed repeatedly with a 5 M solution of HCl and with distilled water until they were free of chlorine ions. Finally, these samples were dried at 200°C for 12 h. The materials were named as KOH:CWZ-22 = X, where X is the mass ratio of KOH:CWZ-22.

2.2. Characterization and adsorption analyses

The textural properties of the ACs were determined by physical adsorption of N_2 at 77 K and CO_2 at 273 K using a Quadrasorb apparatus (Quantachrome Instruments). Before the experiments, the samples were outgassed under vacuum at 250°C overnight. The specific surface area was measured by the multipoint BET (Brunauer–Emmet–Teller) method. The total pore volume, V_p, which includes both the micropores and the mesopores, was estimated from the amount of nitrogen adsorbed at the highest relative pressure. The micropore volume with diameter in the range of 0.31–1.47 nm was calculated by applying the NLDFT model for CO_2 adsorption at 273 K. Based on N_2 adsorption isotherm, the volume of micropores larger than 1.5 nm was obtained by applying the QSDFT method for the slit/cylinder pore model using the software provided by Quantachrome.

ACs were also analyzed by Fourier transform infrared (FTIR) on a Nicollet 380 (Thermo Scientific) spectrometer in order to identify the functional groups on the surface of the ACs. The CWZ-22 carbons were mixed with KBr. FTIR spectra were recorded within a range of 400–4000 cm^{-1}. The equipment was run prior to each actual measurement to record a background spectrum, which was then automatically subtracted from the spectrum of each analyzed sample.

The crystal structures of the samples were determined by XRD with the PANalytical Empyrean X-ray diffractometer using a Cu Kα radiation (α = 1.5418A°) at room temperature.

The surface morphology of the materials was performed by ultra-high resolution field emission scanning electron microscope (UHR FE-SEM Hitachi SU8020) equipped with the secondary electron (SE) detectors.

CO_2 adsorption isotherms were measured up to 2 bar at 25, 40, 60, 80, and 100°C temperature using the volumetric Sieverts' apparatus (IMI—Hiden Isochema Corporation). The CO_2 adsorption data were fitted to standard isotherm models such as Langmuir, Freundlich, Sips, and Toth.

The Langmuir Eq. (1) is based on the assumption of a structurally homogeneous adsorbent and is described by the following [6, 18, 19]:

$$q = \frac{q_m b_L p}{1 + (b_L p)} \tag{1}$$

where q is the adsorbed quantity (mmol of CO_2 per g of AC), p is the pressure of CO_2 in the bulk gas phase, q_m is the maximum adsorption capacity (mmol g^{-1}), and b_L is the Langmuir adsorption affinity constant (bar^{-1}).

The Freundlich isotherm is an empirical equation used to describe heterogeneous systems. The Freundlich Eq. (2) is expressed as [6, 19]:

$$q = k_F p^{n_F} \tag{2}$$

where k_F is the Freundlich constant, and n_F is the Freundlich heterogeneity factor. Value of $n_F > 1$ represents favorable adsorption. The higher n_F value implies higher sorption capacity, such as forming new adsorption sites.

The Sips adsorption isotherm model is a combined form of Langmuir and Freundlich models. The Sips adsorption equation is commonly given by the following nonlinear Eq. (3) [18, 20, 21]:

$$q = \frac{q_m b_s p^{n_s}}{1 + b_s p^{n_s}} \tag{3}$$

where b_s is the Sips adsorption affinity constant, and n_s is the Sips heterogeneity factor.

The Toth isotherm is a model useful in describing heterogeneous adsorption systems, which satisfies both low- and high-end boundaries of the concentration. The Toth's isotherm is expressed as Eq. (4) [20, 22, 23]:

$$q = \frac{q_m b_T p}{[1 + (b_T p)^{n_T}]^{1/n_T}} \tag{4}$$

where b_T is Toth adsorption affinity constant, and n_T is the Toth heterogeneity factor.

Isotherm parameters were obtained through a nonlinear fit of experimental data to the model equations using a MatCad software. In order to evaluate the fit of an isotherm to experimental equilibrium data, algorithms based on the Levenberg–Marquardt were used.

In this study, five nonlinear error functions were examined. In each case, a set of isotherm parameters were determined by minimizing the respective error function across the pressure range studied. The error functions employed were as follows: the correlation coefficient (R^2) [24], the sum of the squares of errors (SSE) [25], the hybrid fractional error function (HYBRID) [25], Marquardt's percent standard deviation (MPSD) [26], and the average relative error (ARE) [25].

Isosteric heats of adsorption (Q_i) can be calculated using Clausius–Clapeyron Eq. (5):

$$\left(\frac{\partial \ln(p)}{\partial \frac{1}{T}} \right)_\theta = \frac{Q_i}{R} \tag{5}$$

where ln(p) is the natural logarithm of the pressure at specific surface loading (θ), Q_i is the isosteric heat of adsorption, R is the universal gas constant ($8.314 \cdot 10^{-3}$ kJ/mol·K), and T is the temperature [K]

The value of isosteric heats of adsorption was determined from the slop of ln(p) versus 1/T.

3. Results and discussion

3.1. Micropore size distribution of activated carbons

Figure 1 shows adsorption–desorption isotherms of N_2 at 77 K for the CWZ-22 and CWZ-22 modified by different amounts of KOH.

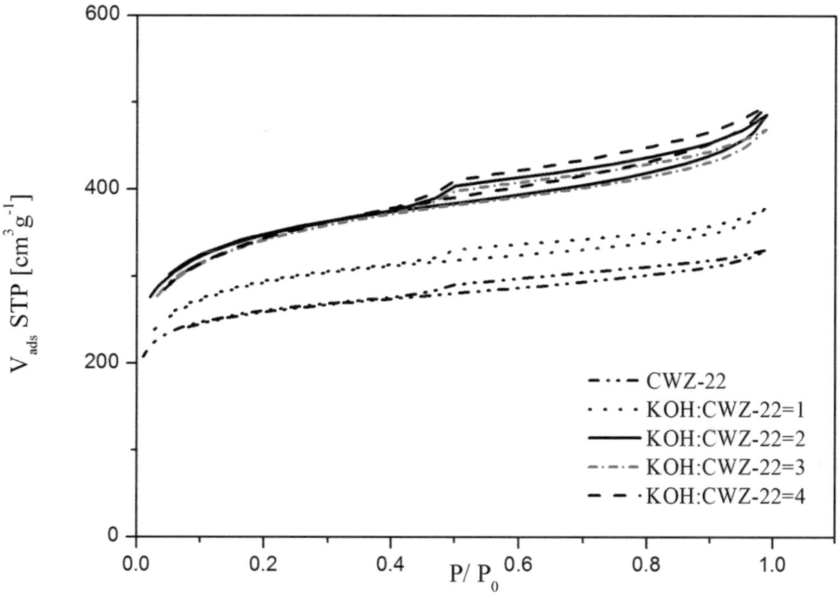

Figure 1. N_2 adsorption–desorption isotherms at 77 K for ACs.

All the isotherms present a high adsorption at low relative pressure, characteristic for microporous materials. The samples from KOH:CWZ-22 = 2–4 ratio showed the higher nitrogen adsorption capacity, whereas for the KOH:CWZ-22 = 1 and CWZ-22, nitrogen adsorption capacity was considerably lower. In addition, all samples showed hysteresis loop at relative pressure >0.4 indicating the presence of certain mesoporosity in the samples. The hysteresis loop type H4 was recognized, which is characteristic of slit shaped pores, such as those present in typical activated carbons. Thus, the isotherms of all samples appear to become a combination of types I and IV according to IUPAC classification. For the isotherms of AC with the highest impregnation ratio, the shape moves toward a type IV isotherm. This can be attributed to a formation of a greater fraction of mesoporosity, which is proved by the mesopore volume (**Table 1**). **Table 1** shows the textural properties of the activated carbons prepared with different KOH:CWZ-22 ratios. Both specific surface area and total pore volume of the samples increased with increasing KOH:CWZ-22 ratios. When the KOH:CWZ-22 ratios are increased from 1 to 2 the specific surface area, total pore volume increases from 1093 to 1292 $m^2 g^{-1}$ and from 0.59 to 0.75 $cm^3 g^{-1}$, respectively. However, further increase of KOH:CWZ-22 ratios only results in slow increase in specific surface area and pore volume. A similar situation can be observed in the case of micropores development. This may be due to the effect of the increase in small pores and the decomposition of porous structure.

AC	S_{BET} [m²/g]	V_p [cm³/g]	V_{mic} (N_2) 1.2–2 nm [cm³/g]	V_{mic} (CO_2) 0.3–1.5 nm [cm³/g]	V_{mes} [cm³/g]
CWZ-22	856	0.51	0.34	0.24	0.17
KOH:CWZ-22 = 1	1093	0.59	0.34	0.31	0.25
KOH:CWZ-22 = 2	1292	0.75	0.38	0.32	0.37
KOH:CWZ-22 = 3	1256	0.73	0.38	0.32	0.35
KOH:CWZ-22 = 4	1299	0.77	0.40	0.37	0.37

Table 1. Textural data obtained by nitrogen (77 K) and CO_2 (273 K) isotherms, using BET and DFT methods.

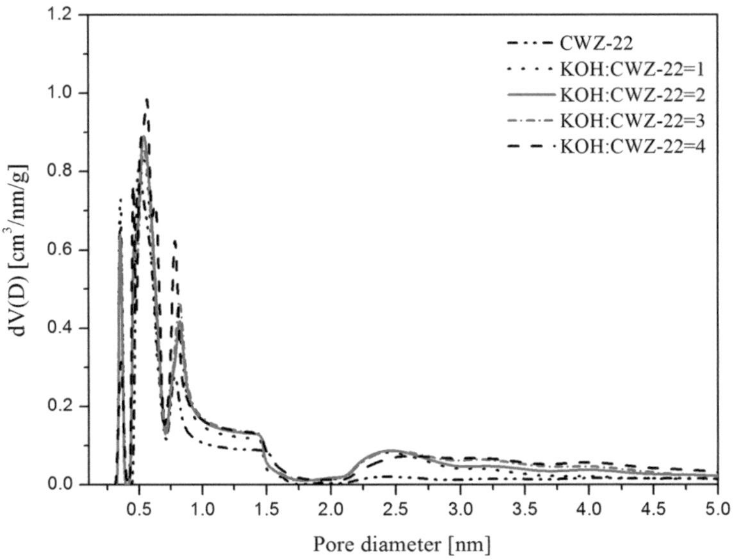

Figure 2. Pore size distribution of the CWZ-22 activated carbons prepared at different KOH:CWZ-22 mass ratio.

CO_2 and N_2 adsorption isotherms are commonly used to determine the pore size distribution of activated carbons. The pore size distribution is shown in **Figure 2**. All activated carbons have peaks in almost the same pore size ranges. One peak is located in the range of 0.3–0.4 nm, the second one in the range of 0.45–0.7 nm, the next in the range of 0.7–0.9 nm, and then in the range of 2.1–4.0 nm. The micropores are dominant. The volume of mesopores (pores

http://dx.doi.org/10.5772/63672

above 2 nm) is low. The pores with a diameter of more than 5 nm were no found. The pore volume of KOH:CWZ-22 = 4 at the pore size of above 0.45–0.7 nm and in the range of 0.7–0.9 nm is larger than for other carbons, indicated on a better developed microporous structure.

3.2. Surface morphology of activated carbon

The SEM images of the pristine CWZ-22 and modified CWZ-22 are shown in **Figure 3**. The surface of CWZ-22 was relatively smooth without large cavities except for some occasional cracks or crevices. All modified CWZ-22 samples have external surfaces covered with irregular holes of different sizes and the irregular shapes. It can be seen from the SEM images that the external surface of the activated carbons was full of cavities. It seems that the cavities on the surface of the carbons resulted from the removal of active agents leaving empty space previously occupied by the active agent. Consequently, the reaction with KOH was aiding in the creation of the porous structure and was proved to be an effective activating agent for the production of high-surface area CWZ-22. Moreover, **Figure 3b–d** shows irregular surface of ACs, the grooves that are considered helpful for the accessibility of gases to the adsorbent surface.

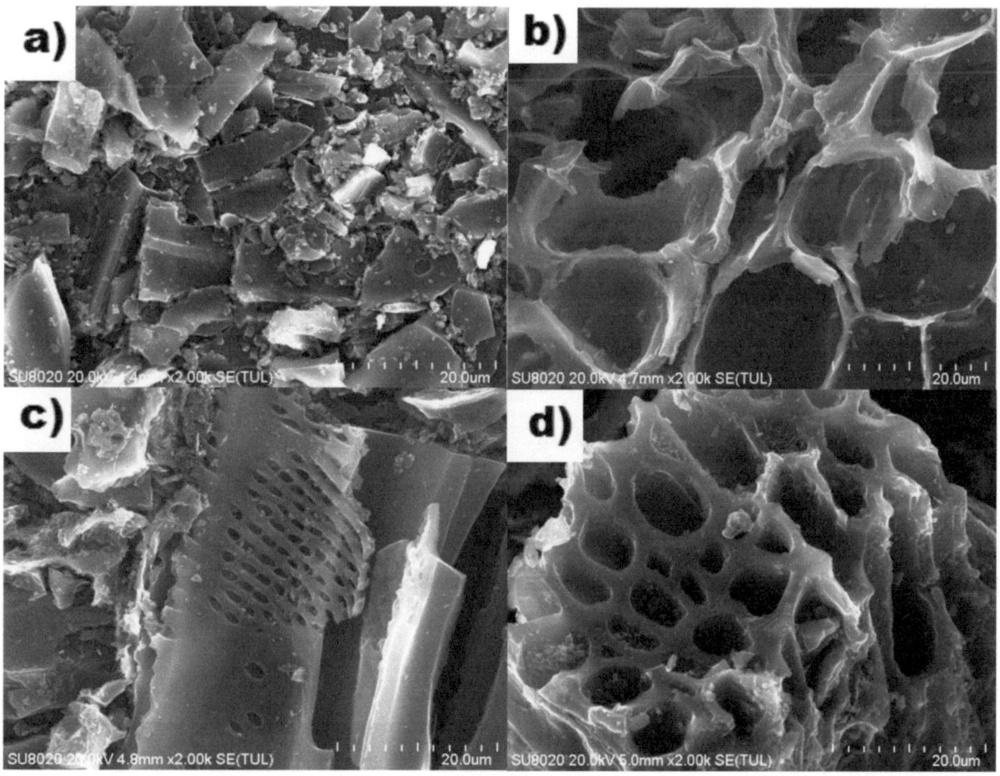

Figure 3. SEM images of the (a) pristine CWZ-22 (b) KOH:CWZ-22 = 1, (c) KOH:CWZ-22 = 2, and (d) KOH:CWZ-22 = 4.

3.3. X-ray diffraction studies

The XRD patterns of pristine CWZ-22 and of the samples with mass ratio from KOH:CWZ-22 = 1 to KOH:CWZ-22 = 4 are shown in **Figure 4**.

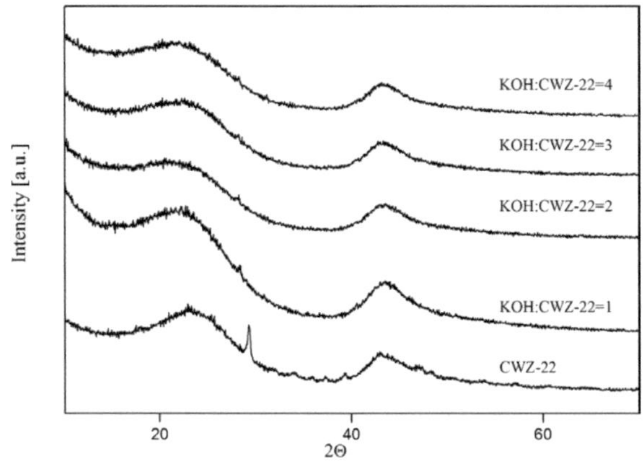

Figure 4. XRD patterns of the ACs.

According to the XRD patterns, the position of peaks is similar for all carbon samples. There are only two peaks at about 23 and 43 2θ in the XRD pattern corresponding to the (002) and (101) planes of the hexagonal graphite (JCPDS, 75-1621). The peaks are very broad because of the random turbostratic stacking of a limited number of layers.

3.4. FTIR analysis

To characterize surface groups on activated carbon, FTIR (Fourier transform infrared) transmission spectra were obtained. The spectra of starting commercial CWZ-22 and modified CWZ-22 were presented in **Figure 5**.

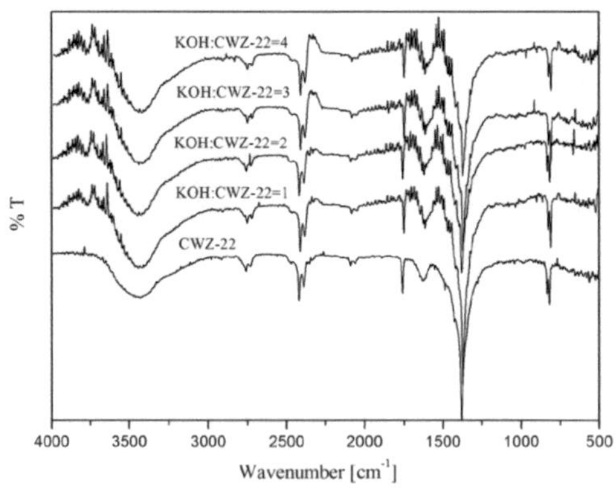

Figure 5. FTIR spectra of activated carbons CWZ-22.

The FTIR spectroscopy provides information on the chemical structure of materials. It can be observed in the **Figure 6** that irrespectively of a KOH/AC ratio, the overall shape of the spectra is very similar. The band at around 3440 cm^{-1} is assigned to the O–H stretching of physisorbed water [27, 28] and that at 1650 cm^{-1}—to a presence of physisorbed water as well as of chemisorbed CO_2. All the spectra show a band in the region 2360–2344 cm^{-1} due to CO_2 in air [28]. The broader peaks about 1760 cm^{-1} are characteristic for C=O stretching vibration in carboxylic groups. The peaks around 1380 cm^{-1} can be assigned to the deformation vibrations of an H–C–OH group [29]. The bands located around 2750 and 820 cm^{-1} could be assigned to C–H group.

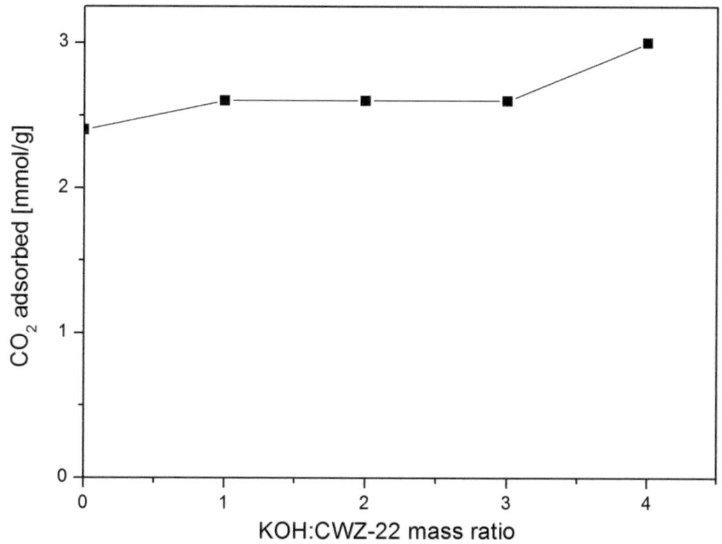

Figure 6. Influence of KOH:CWZ-22 mass ratio on the adsorbed of CO_2 at 1 bar at temperature 25°C.

The FTIR spectra of ACs modified KOH do not exhibit any differences compared to carbon CWZ-22, and these ACs present similar characteristics in their study by the FTIR technique. This suggests that the adsorption capacity of activated carbons CWZ-22 is not depending on the chemical reactivity of functional groups at surface.

3.5. Effect of KOH:CWZ-22 mass ratio of activated carbon on CO_2 adsorption

Preliminary research on CO_2 adsorption was performed at the temperature of 25°C and pressure up to of 1 bar. Chemical activation by KOH increased CO_2 adsorption compared to pristine sample. The difference was small for KOH:CWZ-22 mass ratio 1–3, but when KOH:CWZ-22 was equal to 4, the CO_2 adsorption was considerably higher.

3.6. CO_2 adsorption and modeling of adsorption data

Two ACs samples were chosen for the modeling of adsorption data—the pristine sample and AC with the highest CO_2 adsorption at the temperature of 25°C, namely KOH:CWZ-22 = 4.

The adsorption isotherms were measured at temperatures 25, 40, 60, 80, and 100°C up to the pressure equal to 2 bar. Adsorption isotherms of CO_2 adsorbed on CWZ-22 and KOH:CWZ-22 = 4 at different temperatures are presented in **Figures 7** and **8**, respectively.

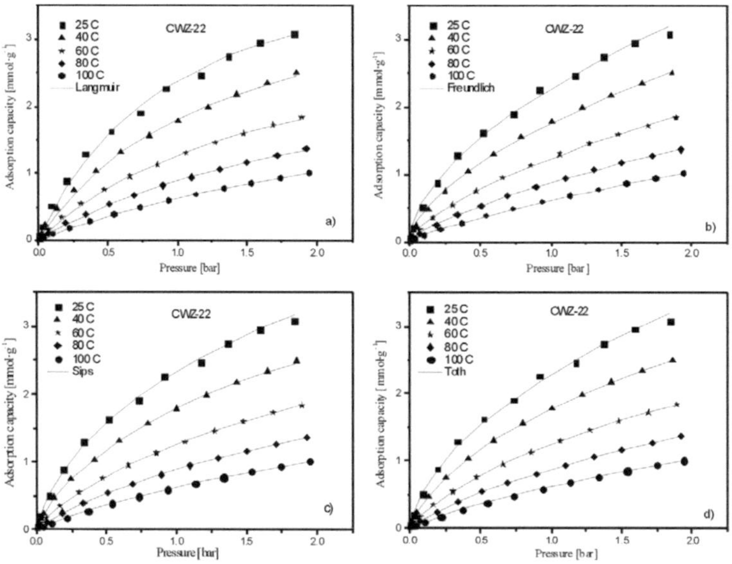

Figure 7. CO_2 adsorption on CWZ-22 in different temperatures. Symbols denote experimental data, and lines correspond to the predictions from the (a) Langmuir, (b) Freundlich, (c) Sips, and (d) Toth isotherm models.

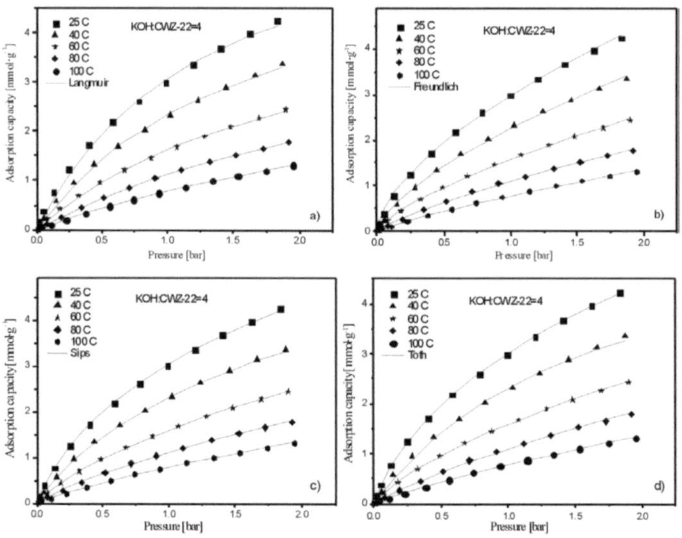

Figure 8. CO_2 adsorption on KOH:CWZ-22 = 4 at different temperatures. Symbols denote experimental data, and lines are predictions from the (a) Langmuir, (b) Freundlich, (c) Sips, and (d) Toth isotherm models.

As it is clearly seen in the **Figures 7** and **8**, the amount of CO_2 uptake increases with an increase in pressure and decreases with increasing adsorption temperature.

The amount of CO_2 adsorbed on the KOH:CWZ-22 = 4 was higher than on the pristine carbon at all temperatures. The adsorption of CO_2 on KOH:CWZ-22 = 4 at 2 bar and 25°C was of 38% higher than on the CWZ-22.

The Langmuir, Freundlich, Toth, and Sips models were employed to describe the adsorption data of CO_2 onto the nonmodified and modified CWZ-22. The fitted adsorption isotherms are presented in **Figures 7** and **8**. Lines show the Langmuir, Freundlich, Toth, and Sips isotherm used to describe adsorption values. The experimental data were marked as symbols. The parameters of model isotherms of Langmuir, Freundlich, Toth, and Sips obtained from a nonlinear fit to experimental data at multiple temperatures for each adsorbate are shown in **Tables 2** and **3**.

Adsorbent	T [°C]	Langmuir			Freundlich		
		q_m [mmol•g^{-1}]	b_L	R^2	k_F [mmol•g^{-1}]	n_F	R^2
CWZ-22	25	4.50	1.13	0.999	2.28	0.55	0.999
	40	4.01	0.84	0.999	1.75	0.60	0.999
	60	3.40	0.60	0.999	1.23	0.66	0.999
	80	2.94	0.44	0.999	0.87	0.71	0.999
	100	2.61	0.32	0.999	0.62	0.76	0.999
KOH:CWZ-22 = 4	25	6.98	0.80	0.999	2.97	0.62	0.999
	40	6.26	0.60	0.999	2.25	0.67	0.999
	60	5.13	0.46	0.999	1.57	0.70	0.999
	80	4.78	0.31	0.999	1.09	0.77	0.999
	100	4.27	0.23	0.999	0.77	0.82	0.999

Table 2. Parameters of Langmuir and Freundlich models for CO_2 adsorption onto the CWZ-22 and KOH:CWZ-22 = 1.

A good agreement can be noticed between the experimental data and the model fitting. This indicates that these models can accurately explain the correlation of the adsorption equilibrium with the adsorbent. According to the R^2 parameter, all the fitted data for the four models are good. In all of the cases, the obtained regression coefficient R^2 is higher than 0.999. Thus, in order to analyze the impact of various error functions on the predicted isotherms, four different error functions namely SSE, HYBRID, MPSD, and ARE were used by nonlinear regression method to recognize the error distribution between the experimental equilibrium data and the isotherms studied. Their values are presented in **Table 4**. The best fit isotherm was selected based on the error function. The lowest values of the error functions are bolded.

Adsorbent	T [°C]	Toth				Sips			
		q_m [mmol•g^{-1}]	b_T	n_T	R^2	q_m [mmol•g^{-1}]	b_S	n_S	R^2
CWZ-22	25	49.71	0.89	0.36	0.999	7.60	0.44	0.74	0.999
	40	28.56	0.13	0.17	0.999	7.78	0.30	0.75	0.999
	60	16.95	0.18	0.32	0.999	6.24	0.25	0.80	0.999
	80	18.74	0.10	0.35	0.999	5.29	0.20	0.84	0.999
	100	14.85	0.08	0.44	0.999	4.41	0.17	0.88	0.999
KOH:CWZ-22 = 4	25	49.60	0.13	0.19	0.999	13.52	0.44	0.76	0.999
	40	43.57	0.28	0.44	0.999	9.83	0.31	0.84	0.999
	60	22.45	0.11	0.35	0.999	9.80	0.26	0.83	0.999
	80	6.60	0.24	0.81	0.999	5.33	0.26	0.97	0.999
	100	6.24	0.16	0.81	0.999	4.95	0.19	0.97	0.999

Table 3. Parameters of Toth and Sips models for CO_2 adsorption onto the CWZ-22 and KOH:CWZ-22 = 1.

	T [°C]	SSE	HYBRID	ARE	MPSD
Langmuir					
CWZ-22	25	0.0621	0.9452	12.2430	24.7905
	40	0.0277	0.5980	8.8268	17.3420
	60	0.0084	0.2504	6.4789	12.7745
	80	0.0024	0.1060	5.9659	12.7569
	100	0.0007	0.0429	3.7235	7.5719
KOH:CWZ-22 = 4	25	0.0730	0.8836	8.3028	15.9486
	40	0.0219	0.2412	4.9931	8.5596
	60	0.0100	0.2578	5.8876	12.0723
	80	0.0005	0.0343	3.0568	8.9733
	100	0.0006	0.0721	3.5310	10.9968
Freundlich					
CWZ-22	25	0.0488	0.9486	12.6447	28.6932
	40	0.0122	0.4071	7.3336	17.5881
	60	0.0054	0.2116	5.9238	13.3919
	80	0.0020	0.1301	6.9945	17.4743
	100	0.0009	0.0575	4.2344	8.9907
KOH:CWZ-22 = 4	25	0.0340	0.7104	7.4561	18.0023
	40	0.0291	1.9369	18.4177	55.9929

	T [°C]	SSE	HYBRID	ARE	MPSD
	60	0.0062	0.1653	4.7272	10.1948
	80	0.0066	0.2695	7.7422	16.0321
	100	0.0025	0.1165	4.5893	9.9664
Sips					
CWZ-22	25	**0.0158**	**0.0827**	**2.0922**	**3.0187**
	40	**0.0003**	**0.0132**	**1.2073**	**3.4548**
	60	**0.0001**	0.0022	**0.2922**	0.9428
	80	**0.0001**	**0.0009**	0.7967	**1.1036**
	100	**0.0002**	**0.0010**	**0.3740**	**0.9455**
KOH:CWZ-22 = 4	25	**0.0014**	**0.0360**	**1.5490**	**4.2991**
	40	0.0028	**0.1062**	**5.7191**	**8.1221**
	60	**0.0001**	**0.0009**	**0.2904**	**0.5330**
	80	**0.0003**	**0.0309**	**2.8569**	**7.7200**
	100	**0.0005**	**0.0678**	**3.5893**	**9.3992**
Toth					
CWZ-22	25	0.0159	0.1261	3.9734	9.8499
	40	0.0022	0.0708	2.9383	7.0922
	60	0.0003	**0.0011**	0.3645	**0.7170**
	80	0.0004	0.0012	**0.6290**	1.6090
	100	0.0003	0.0012	0.5325	1.5068
KOH:CWZ-22 = 4	25	0.0046	0.1036	2.7307	6.7459
	40	**0.0020**	0.2330	5.8074	20.7130
	60	0.0001	0.0040	0.6231	1.8645
	80	0.0004	0.0339	2.9728	8.4997
	100	0.0007	0.0733	3.6244	10.9437

Table 4. The error values of isotherm models for adsorption of CO_2 by CWZ-22 and KOH:CWZ-22 = 4.

The Sips model results in a much better fitting for CO_2 adsorption than Langmuir, Freundlich, and Toth models under the entire experimental temperature range.

Table 3 lists the optimal isotherm parameters obtained by simultaneously fitting of all the adsorption equilibrium data at multiple temperatures for carbons CWZ-22 and KOH:CWZ-22 = 4. The excellent agreement between the model fittings and the experimental data demonstrates that these isotherm models can be confidently employed to accurately correlate the adsorption equilibria of the adsorbates. As it is observed, decreasing the temperature increases the maximum of the adsorbed amount (q_m) in all isotherms. In other words, increasing temperature will decrease the amount adsorbed at a given pressure. The parameter b_s is the adsorption affinity, whereas n_s qualitatively characterizes the heterogeneity of the adsorbate adsorbent system. Also, it can be seen that affinity parameter (b_s) increases with a decrease in temperature. In fact, at lower temperatures, the surface is covered more by the molecules which lead to the stronger affinity of adsorbate molecules towards the surface. The b_s parameters calculated for unmodified and modified carbons are very similar so affinity of CO_2 molecules towards the surface of this two ACs is similar. The surfaces of both ACs were homogenous because n_s was higher than 0.5. The values of n_s for KOH:CWZ-22 = 4 were higher than for the CWZ-22. The modification makes the surface more homogenous. The fitting parameters (qm, b_s, and n_s) of Sips isotherm equations are temperature dependent and defined as Eqs. (6)–(8) [30].

$$q_m = q_{m0} \exp\left[\chi\left(1-\frac{T_0}{T}\right)\right] \tag{6}$$

$$b_s = b_{s0} \exp\left[\frac{Q}{RT_0}\left(\frac{T_0}{T}-1\right)\right] \tag{7}$$

$$n_s = n_{s0} + \alpha\left(1-\frac{T_0}{T}\right) \tag{8}$$

The parameters q_{m0}, b_{s0}, n_{s0} are the parameters q_m, b_s, n_s, respectively, at the same reference temperature (T_0) The $\chi$and α are constant parameters. In Langmuir equation, Q is equal to isosteric heat of adsorption, constant with surface coverage. The parameter Q in the Sips equation is only the measure of the adsorption heat. The parameter Q defined in the affinity constant b_s is the isosteric heat at the fractional loading of 0.5.

	α	Q [kJ/mol]	χ
CWZ-22	0.70	12.08	−2.25
KOH:CWZ-22 = 4	1.10	6.66	−4.91

Table 5. Parameters of Eqs. (6)–(8).

The temperature-dependent equations of the exponent n_s and saturation capacity q_m are empirical, and such forms in Eqs. (6) and (8) were chosen arbitrary because of their simplicity. Because Eqs. (6) and (8) have not been derived from physical laws, then, as a result, the parameters χ and α have no physical meaning.

The unknown parameters of Eqs. (6)–(8) were determined using Excel's Solver to minimize the sum of squared error. The correlated parameters are listed in **Table 5**.

The plots of observed versus predicted values of parameters q_{ms}, b_s, and n_s showed good agreement between these values (**Figure 9**).

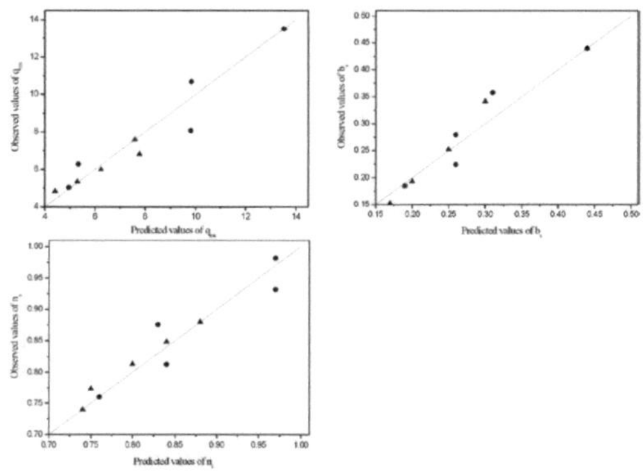

Figure 9. Plot of the observed versus predicted values parameters q_{ms}, b_s, and n_s for CWZ-22 (▲), KOH:CWZ-22 = 4 (●).

On the basis of Eqs. (3), (6)–(8) and values from **Table 3**, A, the adsorption capacity on CWZ-22 and KOH:CWZ-22 = 4 at any temperature can be calculated.

3.7. Isosteric heat of adsorption

The heat of adsorption is described as an indicator of the interaction strength between the adsorbate molecules and the adsorbent. Isosteric heat of adsorption as a function of surface coverage (defined as number of adsorbed molecules on a surface divided by the number of molecules in a filled monolayer on that surface) at different temperatures was calculated by applying the Clausius–Clapeyron equation. The Sips isotherm model was used here for an evaluation of isosteric heat of adsorption. The Q_i of CWZ-22 and KOH:CWZ-22 = 4 was determined from the slope of the straight line after plotting ln(p) against 1/T for different surface coverage (θ), as shown in **Figure 10**.

The isosteric heat of adsorption for CWZ-22 and KOH:CWZ-22 = 4 is shown as a function of surface coverage in **Figure 11**. The isosteric heats of adsorption decreased with the surface

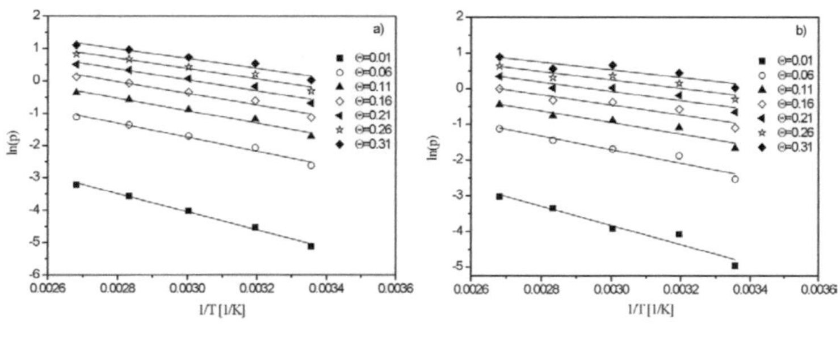

Figure 10. Adsorption isostere for determining isosteric heat of adsorption for ACs: (a) CWZ-22, (b) KOH:CWZ-22 = 4.

coverage increase. For the both ACs at the "zero loading," isosteric heats of adsorption tend to 30 kJ/mol. This value indicates the physisorption. The modification of CWZ-22 reduced the isosteric heat of adsorption. By extrapolation, the curves to surface coverage equal to 0.5 were found that the isosteric heat of adsorption for CWZ and KOH:CWZ-22 = 4 was equal to 10 and 6 kJ/mol, respectively. These values are similar to values obtained on the basis of Eq. (7) and presented in **Table 5**.

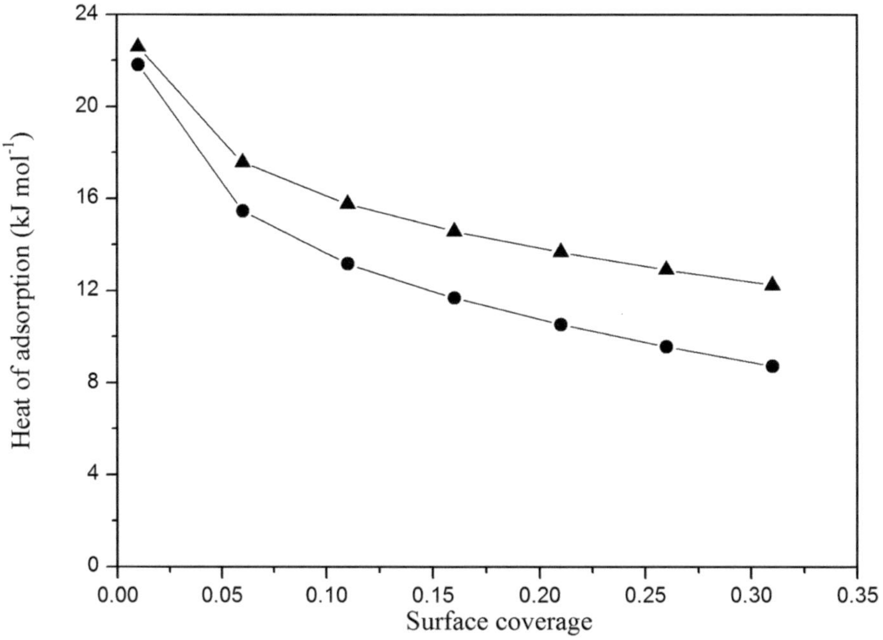

Figure 11. Isosteric heat of adsorption for CO_2 on (▲) CWZ-22 and (–) KOH:CWZ-22 = 4.

4. Conclusions

The impregnation of CWZ-22 by KOH highly effectively increased the porosity, specific surface area, and pore volume of ACs. The specific surface area of KOH:CWZ-22 = 4 significantly increased to 1299 m^2/g as compared with pristine CWZ-22 (856 m^2/g). Similarly, the total pore volume raised from 0.51 to 0.77 cm^3/g.

The chemical modification of CWZ-22 increased the CO_2 adsorption capacity of 38%. The CO_2 adsorption capacity of 4.24 mmol/g obtained for the KOH:CWZ-22 = 4 was considerably higher than that for pristine CWZ-22 (3.08 mmol/g) at 25°C and 2 bar. The Sips model described very good the CO_2 adsorption on KOH:CWZ-22 = 4 and in the pristine sample. The surface of the both ACs was homogenous. The values of isosteric heats of adsorption indicated on physisorption.

Acknowledgements

The research leading to these results has received funding from the Polish-Norwegian Research Programme operated by the National Centre for Research and Development under the Norwegian Financial Mechanism 2009–2014 in the frame of Project Contract No Pol-Nor/237761/98/2014.

Author details

Joanna Sreńscek-Nazzal, Urszula Narkiewicz, Antoni W. Morawski, Rafał J. Wróbel and Beata Michalkiewicz*

*Address all correspondence to: Beata.Michalkiewicz@zut.edu.pl

Institute of Inorganic Technology and Environment Engineering, The West Pomeranian University of Technology in Szczecin, Szczecin, Poland

References

[1] Chen T, Deng S, Wang B, Huang J, Wang Y, Yu G. CO_2 adsorption on crab shell derived activated carbons: contribution of micropores and nitrogen containing groups. RSC Advances. 2015;5:48323–48330. doi:10.1039/C5RA04937G

[2] Deng S, Wei H, Chen T, Wang B, Huang J, Yu G. Superior CO_2 adsorption on pine nut shell-derived activated carbons and the effective micropores at different temperatures. Chemical Engineering Journal. 2014;253:46–54. doi:10.1016/j.cej.2014.04.115

[3] Li P, Tezel FH. Adsorption separation of N_2, O_2, CO_2 and CH_4 gases by beta-zeolite. Microporous and Mesoporous Materials. 2007;98:94–101. doi:10.1016/j.micromeso.2006.08.016

[4] Wu D, Xu Q, Liu D, Zhong C. Exceptional CO_2 capture capability and molecular-level segregation in a Li-modified metal-organic framework. Journal of Physical Chemistry C. 2010;114:16611–16617. doi:10.1021/jp105899t

[5] Linneen N, Pfeffer R, Lin YS. CO_2 capture using particulate silica aerogel immobilized with tetraethylenepentamine. Microporous and Mesoporous Materials. 2013;176:123–131. doi:10.1016/j.micromeso.2013.02.052

[6] Sreńscek-Nazzal J, Narkiewicz U, Morawski AW, Wróbel R, Michalkiewicz B. Comparison of optimized isotherm models and error functions for carbon dioxide adsorption on activated carbon. Journal of Chemical & Engineering Data. 2015;60:3148–3158. doi:10.1021/acs.jced.5b00294

[7] Sreńscek Nazzal J, Glonek K, Młodzik J, Narkiewicz U, Morawski AW, Wróbel R, Michalkiewicz B. Increase the microporosity and CO_2 adsorption of a commercial activated carbon. Applied Mechanics and Materials. 2015;749:17–21. doi:10.4028/www.scientific.net/AMM.749.17

[8] Deng S, Hu B, Chen T, Wang B, Huang J, Wang Y, Yu G. Activated carbons prepared from peanut shell and sunflower seed shell for high CO_2 adsorption. Adsorption. 2015;21:125–133. doi:10.1007/s10450-015-9655-y

[9] Lee SY, Yoo HM, Park SW, Park SH, Oh YS, Rhee KY, Park SJ. Preparation and characterization of pitch-based nanoporous carbons for improving CO_2 capture. Journal of Solid State Chemistry. 2014;215:201–205. doi:10.1016/j.ssc.2014.03.038

[10] Ello AS, Souza LKC, Trokourey A, Jaroniec M. Development of microporous carbons for CO_2 capture by KOH activation of African palm shells. Journal of CO_2 Utilization. 2013;2:35–38. doi:10.1016/j.jcou.2013.07.003

[11] Hu Z, Srinivasan MP. Preparation of high-surface-area activated carbons from coconut shell. Microporous and Mesoporous Materials. 1999;27:11–18. doi:10.1016/s1387-1811(98)00183-8

[12] Olivares-Marín M, Maroto-Valer MM. Preparation of a highly microporous carbon from a carpet material and its application as CO_2 sorbent. Fuel Processing Technology. 2011;92:322–329. doi:10.1016/j.fuproc.2010.09.022

[13] Sevilla M, Fuertes AB. Sustainable porous carbons with a superior performance for CO_2 capture. Energy & Environmental Science. 2011;4:1765–1771. doi:10.1039/C0EE00784F

[14] Shen W, He Y, Zhang S, Li J, Fan W. Yeast-based microporous carbon materials for carbon dioxide capture. ChemSusChem. 2012;5:1274–1279. doi:10.1002/cssc.201100735

[15] Glonek K, Sreńscek-Nazzal J, Narkiewicz U, Morawski AW, Wróbel RJ, Michalkiewicz B. Preparation of activated carbon from beet molasses and TiO_2 as the adsorption of CO_2. Acta Physica Polonica A. 2016;129:158–161. doi:10.12693/APhysPolA.129.158

[16] Mourao PAM, Laginhas C, Custódio F, Nabais JMV, Carrott PJM, Ribeiro Carrott MML. Influence of oxidation process on the adsorption capacity of activated carbons from lignocellulosic precursors. Fuel Processing Technology. 2011;92:241–246. doi:10.1016/j.fuproc.2010.04.013

[17] Heidari A, Younesi H, Rashidi A, Ghoreyshi AA. Adsorptive removal of CO_2 on highly microporous activated carbons prepared from Eucalyptus camaldulensis wood: effect of chemical activation. Journal of the Taiwan Institute of Chemical Engineers. 2014;45:579–588. doi:10.1016/j.jtice.2013.06.007

[18] Ning P, Li F, Yi H, Tang X, Peng J, Li Y, He D, Deng H. Adsorption equilibrium of methane and carbon dioxide on microwave-activated carbon. Separation and Purification Technology. 2012;98:321–326. doi:10.1016/j.seppur.2012.07.001

[19] Hameed BH, Mahmoud DK, Ahmad AL. Equilibrium modeling and kinetic studies on the adsorption of basic dye by a low-cost adsorbent: coconut (*Cocos nucifera*) bunch waste. Journal of Hazardous Materials. 2008;158:65–72. doi:10.1016/j.jhazmat.2008.01.034

[20] Foo KY, Hameed BH. Insights into the modeling of adsorption isotherm systems. Chemical Engineering Journal. 2010;156:2–10. doi:10.1016/j.cej.2009.09.013

[21] Salmasi M, Fatemi S, Doroudian-Rad M, Jadidi F. Study of carbon dioxide and methane equilibrium adsorption on silicoaluminophosphate-34 zeotype and T-type zeolite as adsorbent. International Journal of Environmental Science and Technology. 2013;10:1067–1074. doi:10.1007/s13762-013-0334-9

[22] Ma J, Si C, Li Y, Li R. CO_2 adsorption on zeolite X/activated carbon composites. Adsorption. 2012;18:503–510. doi:10.1007/s10450-012-9440-0

[23] Behvandi A, Tourani S. Equilibrium modeling of carbon dioxide adsorption on zeolites. Chemical and Molecular Engineering. 2011;5:351–353. doi:scholar.waset.org/1999.2/6186

[24] Kumar KV, Porkodi K, Rocha F. Comparison of various error functions in predicting the optimum isotherm by linear and non-linear regression analysis for the sorption of basic red 9 by activated carbon. Journal of Hazardous Materials. 2008;150:158–165. doi:10.1016/j.hazmat.2007.09.020

[25] Porter JF, McKay G, Choy KH. The prediction of sorption from a binary mixture of acidic dyes using single- and mixed-isotherm variants of the ideal adsorbed solute theory. Chemical Engineering Science. 1999;54:5863–5885. doi:10.1016/s0009-2509(99)00178-5

[26] Marquardt DW. An algorithm for least-squares estimation of nonlinear parameters. Journal of the Society for Industrial and Applied Mathematics. 1963;11:431–441.

[27] Yang K, Peng J, Srinivasakannan C, Xia D, Duan X. Preparation of high surface area activated carbon from coconut shells using microwave heating. Bioresource Technology. 2010;101:6163–6169. doi:10.1016/j.biortech.2010.03.001

[28] Murugan C, Bajaj HC, Jasra RV. Transesterification of propylene carbonate by methanol using KF/Al_2O_3 as an efficient base catalyst. Catalysis Letters. 2010;137:224–231. doi:10.1007/s10562-010-0348-6

[29] Li Y, Zijll M, Chiang S, Pan N. KOH modified graphene nanosheets for supercapacitor electrodes. Journal of Power Sources. 2011;196:6003–6006. doi:10.1016/j.jpowsour.2011.02.092

[30] Do DD. Adsorption analysis: equilibria and kinetics. Imperial College Press, London. 1998;57–64.

Chapter 2

Application of Some Natural Porous Raw Materials for Removal of Lead and Zinc from Aqueous Solutions

Mirjana Golomeova and Afrodita Zendelska

Additional information is available at the end of the chapter

http://dx.doi.org/10.5772/62347

Abstract

This chapter discusses the adsorption of lead and zinc ions onto a natural zeolite (clinoptilolite) and zeolite bearing tuff (stilbite) from aqueous solution, whereby the main parameters were the effects of initial pH of solution, mass of adsorbent, and initial metal concentration in solution.

The physical and chemical properties of the natural materials used are characterized by X-ray diffraction, scanning electron microscopy, energy-dispersive spectroscopy, and AES–ICP.

Determination of the maximum capacity of adsorbents for lead and zinc removal under the studied conditions is the main objective of the equilibrium studies. Experimental data are fitted to the Freundlich and Langmuir adsorption models.

The adsorption of lead and zinc ions from diluted solution onto natural zeolite and zeolite bearing tuff occurs efficiently. Determination of the impact of the studied parameters on the efficiency of removal of lead and zinc ions from the solution showed that in the studied range, in the case of lead ions, there were not any significant changes found, but in the case of zinc ions, removal was most effective at lower concentration of zinc ions, higher mass of adsorbents, and higher pH value of the solution.

Keywords: natural zeolite—clinoptilolite, zeolite bearing tuff—stilbite, lead, zinc, adsorption

1. Introduction

Water gets polluted from a number of different sources, and pollutants are divided into various classes, such as organic pollutants, inorganic fertilizers, metals, and radioactive isotopes.

Organic pollutants are susceptible to biological degradation, unlike heavy metals which are not degraded into harmless products [1].

Heavy metals may pose an environmental problem, especially if they migrate into surface water or groundwater. Heavy metals are common pollutants found in various industrial effluents. They are often encountered in metal-plating facilities, electroplating, mining operations, acid mine drainage, fertilizers, battery manufacture, dyestuff, chemical, pharmaceutical, electronic device manufactures, and many others.

Because heavy metals are highly toxic and are nonbiodegradable, they must be removed from the polluted streams in order to meet increasingly stringent environmental quality standards. Their removal can be accomplished by various techniques. Conventional methods typically involve such processes as coagulation, precipitation, ion-exchange, electrochemical methods, membrane processes, extraction, adsorption, etc. Among these, adsorption is currently considered to be more suitable for wastewater treatment, with its high efficiency in heavy metal removal, because of its simplicity and cost- effectiveness. Some widely used adsorbents for adsorption of heavy metals include activated carbon [2], clay minerals [3], biomaterials [4], industrial solid wastes, and zeolites [5–11], characterized by a porous structure. Pore sizes of the materials are defined as follows: macropores, >50 nm; mesopores, 2–50 nm; and micropores, <2 nm.

The adsorption process takes place in three steps: macrotransport, microtransport, and sorption. Macrotransport involves the movement of adsorbate through water to the liquid/solid interface by advection and diffusion. Microtransport involves the diffusion through the macropore system of the solid adsorbent, to the adsorption sites in the micropores and the solid adsorbent. Although adsorption also occurs on the surface of the solid adsorbent and in the macropores and micropores, the surface area of these parts of most solid adsorbents is so small compared with the surface area of the micropores that the amount of material adsorbed there is usually considered negligible.

The respective and widely used adsorbent material in the adsorption processes is zeolite, because it has a high adsorption capacity, surface area, and microporous structure.

1.1. Zeolites

Zeolites are crystalline microporous minerals which are broadly distributed in nature. During the millions of years, the layers of volcanic ash underwent some physical and chemical changes on exposure to high temperatures and pressures, which resulted in the formation of a diverse group of zeolites.

Zeolites have been known for almost 250 years. It was the Swedish mineralogist Axel Fredrick Cronsted, who in 1756, had the privilege to discover the so-called stilbite. He observed that a large amount of steam was obtained upon heating this material. Therefore, this material was named "zeolite," which stems from classical Greek, where ζεω (zeo) means "to boil" and λίθος (lithos) means "stone" [12].

http://dx.doi.org/10.5772/62347

Zeolites are crystalline aluminosilicates with open 3D framework structures built of SiO_4 and AlO_4 tetrahedra linked to each other by sharing all the oxygen atoms to form regular intracrystalline cavities and channels of molecular dimensions. Zeolite frameworks are made up of four coordinated atoms forming tetrahedra, which are linked by their corners. This feature makes a rich variety of beautiful structures of zeolite. The framework structure of zeolite contains channels, cages, and cavities. These are linked and big enough to allow easy drift of the resident ions and molecules into and out of the structure. Zeolite's low specific density is the result of the system of large voids, which are interconnected and form long wide channels of various sizes depending on the compound [13]. This ability puts zeolites in the class of materials known as "molecular sieves."

A description of zeolite structure usually begins with a description of the framework type in terms of the size of the pore openings and the dimensionality of the channel system. Pore opening is characterized by the size of the ring that defines the pore, designated an *n*-ring, where *n* is the number of T-atoms in the ring, and T is any tetrahedrally coordinated cation (aluminum or silicon atom). An 8-ring is considered to be a small pore opening, a 10-ring a medium one, and a 12-ring a large one, with free diameter (calculated using an oxygen radius of 1.35 Å) of approximately 4.1, 5.5 and 7.4 Å, respectively. Of course, rings can be distorted considerably; so, these numbers should only be used as a rough guide [14]. For zeolites containing larger rings, ions and molecules can enter the intracrystalline space. For rings that contain 6 T atoms (six-membered rings or 6-ring) or less, the size of the window is ~2 Å, and movement of species through these rings is restricted. Ions or molecules can be trapped in cages bound by rings of this size or smaller (5-, 4-, or 3-ring) [15].

The internal volume of zeolites consists of interconnected cages or channels, which can have dimensionalities of one or three. Pore sizes can vary from 0.2 to 0.8 nm, and pore volumes from 0.10 to 0.35 cm^3/g [15].

Another unique property of zeolites is that they are one of the few "negatively charged" minerals found in nature. The aluminosilicate framework is negatively charged and attracts the positive cations that reside in cages, such as sodium (Na^+), potassium (K^+), calcium (Ca^{+2}), and magnesium (Mg^{+2}), to compensate the negative charge of the framework. Unlike most other tectosilicates, zeolites have larger cages in their structure [13].

Zeolites are also characterized by the unique property that the internal surface is highly accessible and can compose more than 98% of the total surface area. Surface area is typically of the order of 300–700 m^2/g [15].

Thermal stability of zeolites varies over a large temperature range. The decomposition temperature for low-silica zeolites (Si/Al ratio ≈ 1) is ~700°C, whereas completely siliceous zeolite, such as silicalite, is stable up to 1300 °C. Low-silica zeolites are unstable in acid, whereas high-silica zeolites (Si/Al ratio ≥ 10) are stable in boiling mineral acids, though unstable is basic solutions. Low-silica zeolites tend to have structures with 4, 6, and 8 rings, whereas more siliceous zeolites contain 5-ring structure. Low-silica zeolites are hydrophilic, whereas high-silica zeolites are hydrophobic [15].

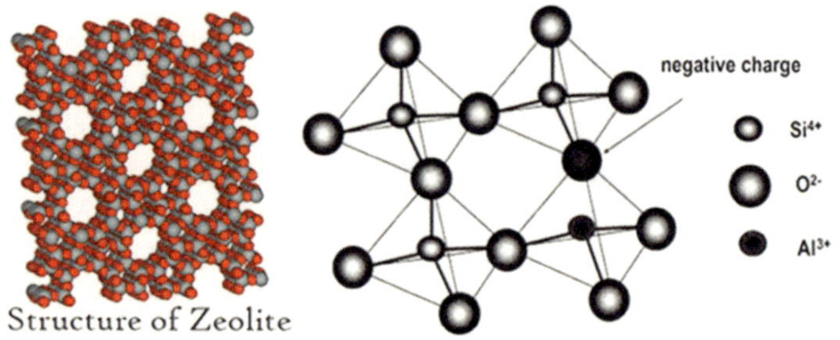

Figure 1. Structure of zeolite.

The properties of the porous materials depend both on the chemistry of the framework and the pore structures (**Figure 1**). The growing need for materials with highly specific physical and chemical properties as zeolites has inspired scientists to make new materials with high porosity and unique structures. Many of the zeolites occur naturally as minerals, but most of them have been made synthetically; some of them have been made for commercial use, while others have been created by scientists to study their chemistry. At present, there are 191 unique zeolite frameworks identified [16], and over 40 naturally occurring mineral frameworks are recognized members of the zeolite group. These are the members of the zeolite group: Analcime Family, Chabazite Family, Gismondine Family, Harmotome Family, Heulandite Family, Natrolite Family, and Stilbite Family.

Zeolites are quite complex structures; hence, their classification requires several attempts. However, the old classification of zeolites does not lose its importance, and forms the basis of morphological significance of zeolites. All zeolites can be classified into three groups: fibrous zeolites, platy or lamellar zeolites, and cubic or robust zeolites. This classification was later expanded in order to include other natural zeolites. Instead of physical properties, classification was based on the presence of a complex structural unit of tetrahedron that was considered to constitute a fundamental unit characteristic of each group or family. Some minor variations in these units were considered acceptable, allowing more zeolite topologies to be categorized into only a handful of groups. The following categories of zeolites were reported: chain or fibrous zeolites, singly connected 4-ring chains, doubly connected 4-ring chains, 6-ring zeolites, hexagonal sheets with handles, and the heulandite group which included heulandite, clinoptilolite, stilbite and its variants, and brewsterite [17].

Zeolites are very useful minerals. They have been used in various industries recently. This is due to their many attractive characteristics. There are three main uses of zeolites in industry: catalysis, adsorption, and ion exchange.

Zeolites have been used extensively as catalysts. The microporous nature gives zeolite a high surface area where reactions can take place. Because of the shape-selective nature of zeolite, they are sometimes referred to as shape-selective catalysts [10]. The main industrial application areas are: petroleum refining, fuel production, and petrochemical production. Synthetic zeolites are the most important catalysts in petrochemical refineries.

http://dx.doi.org/10.5772/62347

Zeolite contains water molecules and cations; when water is removed from zeolite, empty voids are created within its framework, which can be occupied by other molecules. The occupation of these voids by other molecules is called adsorption [10]. Zeolites are used to adsorb various materials, because of which they can be of use in drying, purification, and separation processes. A widely used property of zeolites is that of gas separation. Other applications that can take place within the pore include polymerization of semiconducting materials and conducting polymers to produce materials having unusual physical and electrical attributes.

The zeolite structure contains exchangeable cations (K^+, Na^+, Ca^{+2}, Mg^{+2}), which are readily exchanged with other types of cations from solution. The alkali metals such as Na or K can be moved out of the zeolite and can be replaced by the "hard" Ca and Mg ions from the water. This property of zeolite has been exploited in a major way in water softening.

Application of zeolite has been found to be attractive in the removal of heavy metals from waste water. In agronomy, horticulture, and soil remediation zeolites can be added to increase fertilizer efficiency and to reduce the leaching of nutrients. Zeolites can also be added as animal feed supplements. Their molecular sieve properties are used to trap or separate gases in agriculture (e.g. ammonia) and to treat effluents containing radioactive contaminants or other heavy metals.

Application of zeolites in waste water treatment is very important. The properties that make natural zeolite an attractive alternative for the treatment of waste water are as follows: They are cheap since they are relatively abundant [18]. They have a favorable cation exchange capacity (CEC) [19]. They have good selectivity for cations [20]. Zeolites have a high surface area due to their porous and rigid structure [21]. They also act as molecular sieves, and this property can easily be modified to increase the performance of the zeolite [22]. In acidic conditions, the zeolites have good structural stability and can be regenerated easily. The acidic solutions can be neutralized by zeolite, which is achieved through the exchange of H^+ ions from solution with the cations in the zeolite structure [23].

Because of these features, zeolites have been a growing interest in the adsorption of heavy metals from solution. Many researchers have shown the feasibility of using natural zeolite in the adsorption of heavy metals under different experimental conditions such as temperature, pH, concentration, and agitation speed [8, 10, 24].

1.1.1. Clinoptilolite

Clinoptilolite belong to clinoptilolite series minerals and is the most common zeolite.

These minerals occur in rocks and sediments formed in continental accumulations, deep sea sediments, and in some lava-flow sections [25].

Clinoptilolite occur as fine-grained crystals in massive beds, and important deposits of clinoptilolite occur in several countries: Bulgaria, Mexico, Hungary, Cuba, Italy, Jordan, United States, and Russia.

The origin and use of the name clinoptilolite has a convoluted history, from 1890. The clinoptilolite series comprises three species: Clinoptilolite-K (Clinoptilolite with K-dominant crystals—$(K,Na,Ca_{0.5},Sr_{0.5},Ba_{0.5},Mg_{0.5})_6(H_2O)_{20} | [Al_6 Si_{30} O_{72}]$) is the name for the original material from the ridge east of Hoodoo Peak, Wyoming; Clinoptilolite-Na ($(Na,K,Ca_{0.5},Sr_{0.5},Ba_{0.5},Mg_{0.5})_6 (H_2O)_{20} | [Al_6 Si_{30} O_{72}]$) is the name for Na-dominant crystals with the suggested type example from the Barstow formation, San Bernardino County, California, USA; Clinoptilolite-Ca ($(Ca_{0.5},Na,K,Sr_{0.5},Ba_{0.5},Mg_{0.5})_6(H_2O)_{20} | [Al_6 Si_{30} O_{72}]$) for Ca-dominant samples with type examples from Kuruma Pass, Fukus hima Prefecture, Japan.

Clinoptilolite has the same tetrahedral framework like that of heulandite [16] and form a continuous compositional series, sometimes referred to as the heulandite group zeolites. The crystal structures of clinoptilolite and heulandite are mostly described to be monoclinic, space group *C2/m*, but lower symmetries such as *Cm* and C1 have also been reported [16]. The framework contains three sets of intersecting channels all located in the (010) plane. Two of the channels are parallel to the *c*-axis: the A channels are formed by strongly compressed 10-membered rings (aperture 3.0 × 7.6 Å), and B channels are confined by 8-membered rings (aperture 3.3 × 4.6 Å). C channels are parallel to the *a*-axis, or [102] and are also formed by 8-membered rings (aperture 2.6 × 4.7 Å) (**Figure 2**) [26].

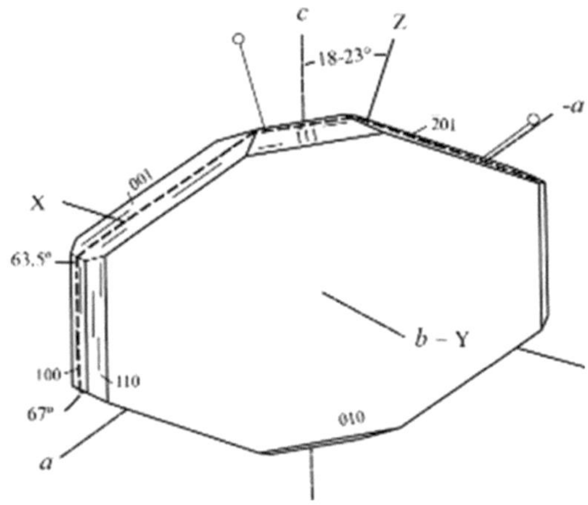

Figure 2. Crystallography of clinoptilolite.

The other characteristics of clinoptilolite are as follows: It is colorless, white, pink, yellow, reddish, or pale brown. Crystals are transparent to translucent. Cleavage is perfect in one direction, parallel to the prominent pinacoid face. Fracture is uneven. Hardness is 3.5–4, maybe softer on cleavage surfaces. Specific gravity is approximately 2.2 (very light).

1.1.2. Stilbite

The name, stilbite, is derived from the Greek word for mirror in allusion to its luster on the cleavage plane. The occurrences of stilbite are mentioned since 1801 by Haüy. According to Coombs [27], stilbite-Ca is the name for the original material, in which Ca is the most abundant nonframework cation and stilbite-Na is a species with the type example from Capo Pula, Cagliari, Sardinia, Italy [28].

Stilbite-Ca is a common zeolite, while stilbite-Na is rare. Stilbite-Ca occurs in fractures and other cavities in basaltic rocks. Stilbite-Ca occurs as a vein mineral in diagenetically altered or metamorphosed volcaniclastic rocks. It also occurs as a vein mineral cutting nonvolcanic rocks, such as pegmatite, gneiss, schist, or granite.

The framework type [16, 28] of the stilbite group, which includes stilbite series, stellerite, and barrerite, consists of two sets of connected channels. One channel extends parallel to the *a*-axis and is confined by a 10-membered ring (aperture 4.9 × 6.1 Å). The other channel (aperture 2.7 × 5.6 Å) is located along [101] for monoclinic frameworks or along [001] for orthorhombic structure, and is confined by an 8-membered ring. Both of these channels are in the (010) plane, creating a structural weakness across the plane leading to perfect (010) cleavage and a tabular habit (**Figure 3**).

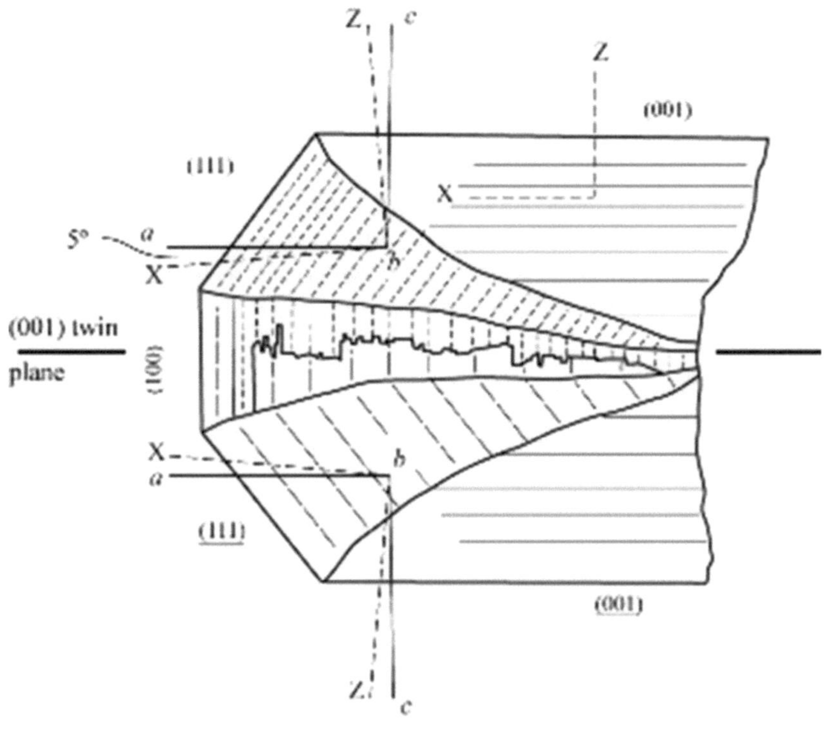

Figure 3. Crystallography of stilbite.

The other characteristics of stilbite are: the color is pink or white, and also tinted yellow and red; luster is vitreous to pearly, especially on the prominent pinacoid and cleavage surfaces; crystals are transparent to mostly translucent; cleavage is perfect in one direction, parallel to the prominent pinacoid; fracture is uneven; hardness is 3.5–4; and specific gravity is approximately 2.2 (very light).

1.2. Heavy metals

1.2.1. Lead

Lead is a heavy, low-melting, bluish-gray metal that occurs naturally in the Earth's crust. However, it is rarely found naturally as a metal. It is usually found combined with two or more other elements to form lead compounds.

Lead can enter the environment through releases from mining lead and other metals, and from factories that make or use lead, lead alloys, or lead compounds. Once lead gets into the atmosphere, it may travel long distances, if the lead particles are very small. Lead is removed from the air by rain and by particles falling to land or into surface water, but in small amounts. Lead is commonly found in soil, especially near roadways, mining areas, industrial sites, near power plants, incinerators, landfills, and hazardous waste sites. Sources of lead in surface water or sediments include deposits of lead-containing dust from the atmosphere, waste water from industries that handle lead (primarily iron and steel industries and lead producers), urban runoff, and mining piles.

People who live and work near hazardous waste sites may be exposed to lead by breathing air, drinking water, eating foods, or swallowing dust or dirt that contain lead. Once this lead gets into the lungs, it travels quickly to other parts of the body in the blood and organs (such as the liver, kidneys, lungs, brain, spleen, muscles, and heart). After several weeks, most of the lead moves into the bones and teeth and stay there for decades. The main target for lead toxicity is the nervous system. Long-term exposure results in decreased functions of the nervous system. Exposure to lead can cause weakness in fingers, wrists, or ankles, small increases in blood pressure, anemia, and severe damage to the brain and kidneys [29].

1.2.2. Zinc

Zinc is one of the most common elements in the Earth's crust. Zinc is present in air, soil, and water, and in all foods.

Zinc can enter the environment through releases from mining, purifying of zinc, lead, and cadmium ores, steel production, coal burning, burning of wastes, zinc chemical industries, and domestic waste water. Sludge and fertilizers also contribute to increased concentration of zinc in the soil. All these sources can increase the concentration of zinc in the atmosphere, soil, and water. In air, zinc is present mostly as fine dust particles, which settle over land and water.

Zinc is an essential element needed by human body in small amounts. Humans are exposed to zinc compounds in food. Zinc can enter the body through the digestive tract when you eat food or drink water containing it. Zinc can also enter through lungs if zinc dust or fumes from zinc-smelting or zinc-welding operations is inhaled. Zinc is stored throughout the body. The concentration of zinc increases in blood and bone rapidly after exposure, may stay in the bone for many days, and leave the body in urine and feces. Inhaling large amounts of zinc (as zinc dust or fumes from smelting or welding) can cause a specific short-term disease called metal fume fever, which is generally reversible once exposure to zinc ceases [30].

http://dx.doi.org/10.5772/62347

2. Characteristics of materials

Particle characterization reveals information on the physical and chemical nature of natural zeolite particles, which is related to its ability to remove heavy metal ions from solution.

Two types of natural raw materials were used in a recent study for removal of heavy metals such as lead and zinc. The first of them was natural zeolite from Kardjali, Republic of Bulgaria. The second natural raw material was zeolite bearing tuff from Vetunica deposits, localized in the northern marginal parts of the well-known Kratovo-Zletovo volcanic area in the Republic of Macedonia. The particle size range of the natural raw materials used in this study was 0.8–2.5 mm.

The general characteristics of used materials, such as chemical composition, physical characteristics, and cation exchange capacity are presented in **Tables 1**, **2**, and **3**, respectively.

Chemical composition (%)	Natural zeolite	Zeolite bearing tuff
SiO_2	69.68	54.67
Al_2O_3	11.40	20.16
Fe_2O_3	0.93	–
CaO	2.01	4.86
MgO	0.87	1.08
K_2O	2.90	2.40
TiO_2	0.15	0.45
Na_2O	0.62	1.97
MnO	0.08	0.06
H_2O	13.24	–
P_2O_5	0.02	0.24
NaO	–	–
FeO	–	3.98

Table 1. Chemical composition of natural raw materials.

	Hydrated density (g/cm^3)	Dehydrated density (g/cm^3)	Porosity (%)
Natural zeolite	2.31	1.84	20.20
Zeolite bearing tuff	1.72	0.89	48.40

Table 2. Density and porosity of natural raw materials.

Chemical composition (%)	Natural zeolite	Zeolite bearing tuff
K^+	41 mEq/100g	66.5 mEq/100g
Na^+	16.10 mEq/100g	3.5 mEq/100g
Ca^{2+}	67.14 mEq/100g	21.5 mEq/100g
Mg^{2+}	3.88 mEq/100g	8.5 mEq/100g
Total cation exchange capacity	1.8–2.2 mEq/g	0.69–1.07 mEq/g

Table 3. Cation exchange capacity.

X-Ray Diffractometer 6100 from Shimadzu was used to investigate the mineralogical structure of natural raw material samples. This technique is based on observing the scattering intensity of an X-ray beam hitting a sample as a function of incident and scattered angle, polarization, and wavelength or energy. The diffraction data obtained are compared with the database maintained by the *International Centre for Diffraction Data,* in order to identify the material in the solid samples. The results of XRD are presented in **Figures 4** and **5**.

Figure 4 presented the sample taken from Kardjali, Bulgaria, and showed that natural zeolite contained majority (90–94%) of clinoptilolite.

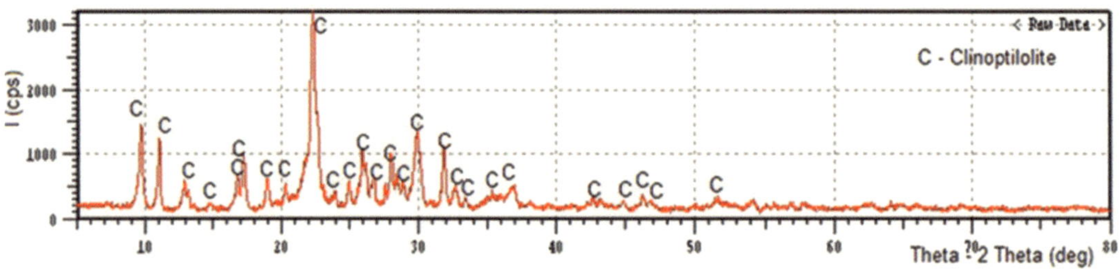

Figure 4. X-ray diffraction of a sample from Kardjali, Bulgaria.

The second sample is taken from Vetunica deposit, Kratovo, Macedonia, and is presented in **Figure 5**. The result showed that this sample contained about 27% stilbite ($NaCa_2Al_5Si_{13}O_{36} \cdot 14H_2O$) and that the rest were albite, anorthite, kaolinite, and quartz. This analysis confirms the results obtained from the study of Blazev [31], where Diffractometer PHILIPS, type PW 1051, was used for the determination of the mineralogical composition.

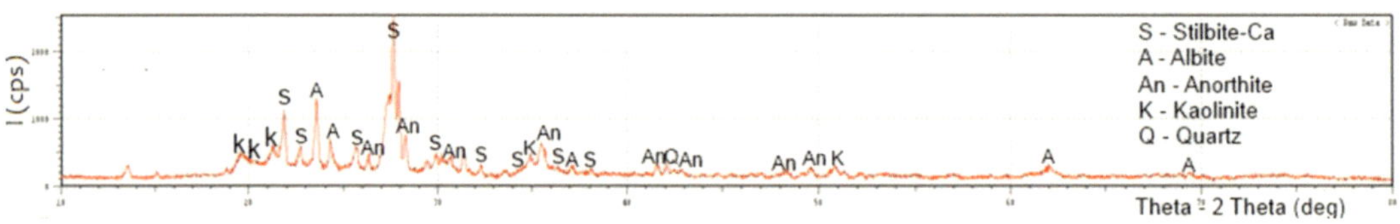

Figure 5. X-ray diffraction of sample from Vetunica, Kratovo, Macedonia.

http://dx.doi.org/10.5772/62347

Scanning electron microscope, VEGA3 LMU, fitted with an Inca 250 EDS (Energy-Dispersive Spectroscopy) system, is used to study the surface morphology of natural raw material samples. The elemental analysis of a sample is based on the emission of characteristic X-rays by the sample when subjected to a high-energy beam of charged particles such as electrons or protons. Micrographs of natural raw material samples are shown in **Figures 6** and **7**. The micrographs clearly show numerous macropores and well-defined crystals of clinoptilolite and stilbite in the zeolite structure in both the samples.

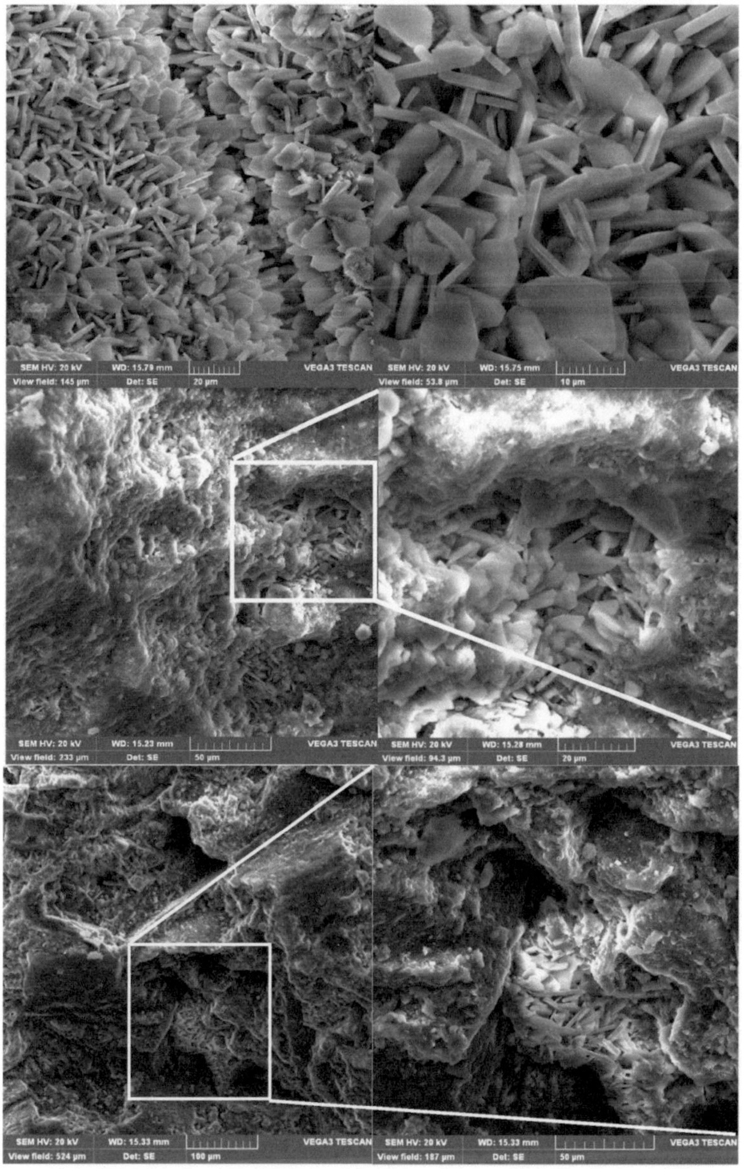

Figure 6. Micrographs of natural zeolite samples obtained from SEM analysis.

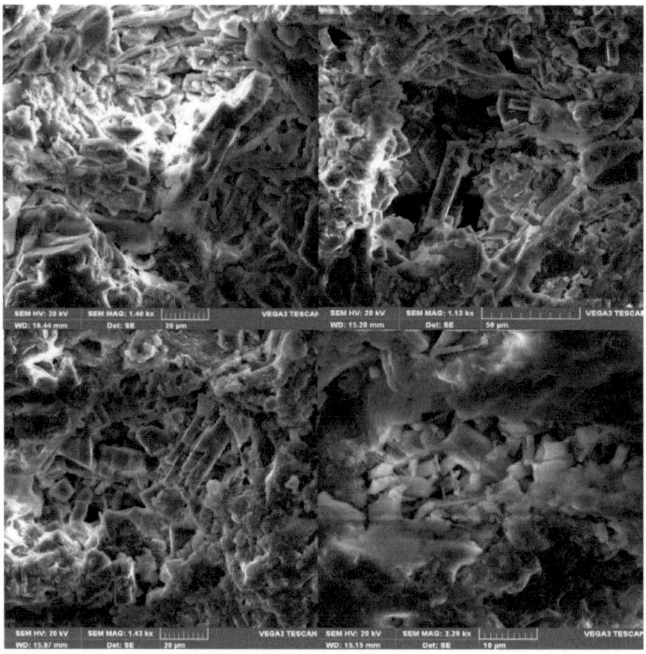

Figure 7. Micrographs of zeolite bearing tuff samples obtained from SEM analysis.

An electron beam was directed onto different parts of the samples in order to get a more accurate analysis (**Figure 8**). Obtained elemental compositions of natural zeolite-—clinoptilolite and zeolite bearing tuff—stilbite are presented in **Table 4**.

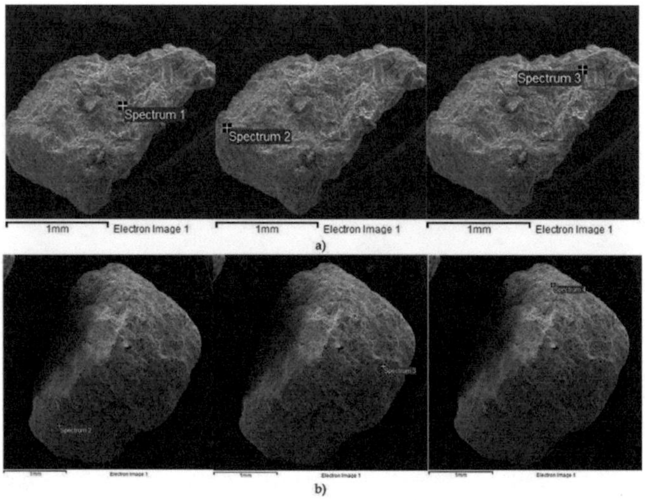

Figure 8. EDS analysis showing the scanning method for natural raw material samples: (a) Natural zeolite—clinoptilolite and (b)Zeolite bearing tuff—stilbite.

Element	Natural zeolite—Clinoptilolite	Zeolite bearing tuff—Stilbite
	Average (%)	
O	57.56	61.62
Na	0.24	0.34
Mg	0.72	1.07
Al	5.28	8.13
Si	30.13	26.29
K	2.71	1.65
Ca	1.99	0.91
Fe	1.38	–
Total	100	100

Table 4. EDS analysis showing the elemental composition for natural raw material samples.

Results of EDS analysis showed that the predominant exchangeable cations in natural zeolite—clinoptilolite structure were K^+ and Ca^{2+}, while the predominant exchangeable cation in stilbite rich tuff structure was K^+, which was then followed by Mg^{2+} and Ca^{2+}.

3. Effect of studied conditions

The rate of adsorption is a complex function of several factors such as initial pH and concentration of solution, mass of adsorbent, adsorbent particle size, temperature, flow rate in columns, and agitation speed in the case of batch experiments. The overall reaction rate may be influenced by separate or combined effect of these factors. In these studies, some of these factors are investigated with regard to their effect on the efficiency of natural zeolite and zeolite bearing tuff in removing lead and zinc from solutions.

For this purpose, adsorption of lead and zinc ions on natural porous mineral materials was studied with synthetic single-ion solutions of lead and zinc with different initial concentrations (5, 25, 50, 200, and 400 mg/l). Synthetic single-component solutions of these metals were prepared by dissolving a weighed mass of the analytical grade salt $Pb(NO_3)_2$ and $ZnSO_4 \cdot 7H_2O$, appropriately, in 1000 ml distilled water.

Initial pH of the prepared solution was adjusted by adding 2% sulfuric acid and controlled by 210 Microprocessor pH meter. Initial pH ranges of tested solution were 2.5, 3.5, and 4.5. The experiments were performed in a batch mode in a series of beakers equipped with magnetic stirrers contacting a mass of adsorbents (2, 5, and 10 g) with a certain volume of 400 ml of metal ion solution. Magnetic stirrer at 400 rpm was used for agitation up to 360 min, at room temperature of 20 ± 1°C. The final pH value was also measured. After predetermined time, the suspension was filtered, and the filtrate was analyzed using Liberty 110, ICP Emission Spectrometer, Varian. Inductively coupled plasma atomic emission spectroscopy (ICP-AES)

Agilent was used to analyze the concentration of metal ions in solution. Inductively coupled plasma atomic emission spectroscopy (ICP-AES) is an analytical technique used for the detection of trace metals. It is a type of emission spectroscopy that uses the inductively coupled plasma to produce excited atoms and ions that emit electromagnetic radiation at wavelengths characteristic of a particular element. The intensity of this emission is indicative of the concentration of the element within the sample [46, 47].

Based on material balance, the adsorption capacity was calculated by using the following expression [32]:

$$q_e = \frac{V(C_0 - C_e)}{m} (\text{mg/g}) \quad (1)$$

where q_e is the mass of adsorbed metal ions per unit mass of adsorbent (mg/g), C_0 and C_e are the initial and final metal ion concentrations (mg/dm^3), respectively, V is the volume of the aqueous phase (l), and m is the mass of adsorbent used (g).

$$AD\% = \left(1 - \frac{C_e}{C_0}\right) \cdot 100 \quad (2)$$

The difference in adsorption capacity of the adsorbents for the heavy metal ions may be due to numerous factors, which include hydration radii, hydration enthalpies, and solubility of the cations. The hydration radii of the studied cations are: $r_H Zn^{2+} = 4.30$Å and $r_H Pb^{2+} = 4.01$Å [33, 34]. The smallest cations should ideally be adsorbed faster and in larger quantities compared to the larger cations, since the smaller cations can pass through the micropores and channels of the zeolite structure with ease [8]. Furthermore, adsorption should be described using hydration enthalpy, which is the energy that permits the detachment of water molecules from cations and thus reflects the ease with which the cation interacts with the adsorbent. Therefore, the more a cation is hydrated, the stronger its hydration enthalpy and the less it can interact with the adsorbent [10]. Divalent cations with low hydration energies are adsorbed preferably compared to cations with high hydration energies [35]. The hydration energies of the studied cations are: −1955 and −1481 kJmol^{-1} for Zn^{2+} and Pb^{2+}, respectively [33, 34].

The lead ion has smaller hydration radius and hydration energy than zinc ion. Based on this fact it is expected that adsorption of lead ions will be better than zinc ions.

3.1. Effect of initial metal concentration in solution

The effect of initial concentration was investigated by contacting 5 g of adsorbent, at pH 3.5, with different concentrations of single-component solutions, 5, 25, 50, 200, and 400 mg/l. An increase in concentration generally results in an increase in the amount of heavy metals adsorbed. This may be due to an increase in the number of collisions between the reactants, leading to the observed increase in reaction rate and capacity, according to the collision theory

[36]. Increasing the initial metal concentration in solution until the system reaches a saturation point will result in increase of adsorption capacity. After reaching a saturation point, increasing the adsorbate (lead and zinc) concentration will not result in any significant change in the amount adsorbed, q_e. Results of this investigation are presented in **Table 5**.

Heavy metal ion	Adsorbent	Initial concentration C_o (mg/l)	Amount adsorbed q_e (mg/g)	Percentage adsorbed (%)
Lead	Natural zeolite—clinoptilolite	5	0.390	97.60
		25	1.984	99.20
		50	3.982	99.54
		200	15.642	97.76
		400	30.143	94.19
	Zeolite bearing tuff—stilbite	5	0.400	99.98
		25	2.000	100
		50	4.000	100
		200	15.500	96.9
		400	27.224	85.1
Zinc	Natural zeolite—clinoptilolite	5	0.376	94
		25	1.786	89.32
		50	2.901	72.52
		200	3.619	26.62
		400	7.263	26.69
	Zeolite bearing tuff—stilbite	5	0.3896	97.4
		25	1.937	96.84
		50	3.632	90.8
		200	13.152	82.2
		400	25.408	79.4

Table 5. Effect of initial solution concentration on the adsorption capacity of natural zeolite and zeolite bearing tuff.

The above results indicate that the amount of lead and zinc adsorbed by natural zeolite and zeolite bearing tuff at equilibrium is dependent on the initial metal concentration in solution. As initial concentration increases, the adsorption of lead and zinc ions by natural zeolite and zeolite bearing tuff also increases. This was expected because it is a consequence of an increase in the concentration-driving force. The concentration-driving force is important because it is responsible for overcoming the mass transfer resistance associated with the adsorption of metals from solution by the adsorbent [37].

Increase in initial metal concentration in solution not only results in an increase in the amount adsorbed (q_e) but also a decrease in the efficiency of adsorbents used for the removal of lead and zinc from solution. This conclusion can be seen from **Table 5**. Sprynskyy et al. [22] and Motsi [10] also found a similar trend, that is, a decrease in efficiency, in their work, on the adsorption of lead, copper, nickel, and cadmium from solution by clinoptilolite [22] and on the adsorption of iron, copper, manganese, and zinc from solution by clinoptilolite [10].

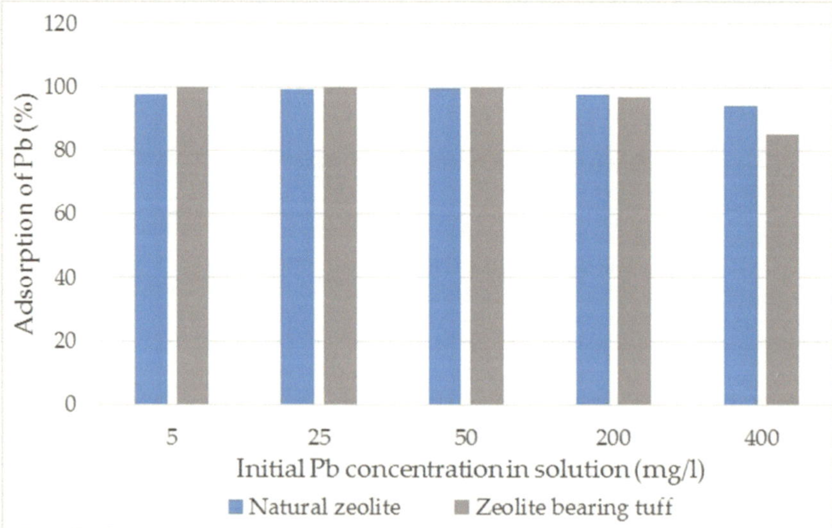

Figure 9. Adsorption of lead depending on initial metal concentration.

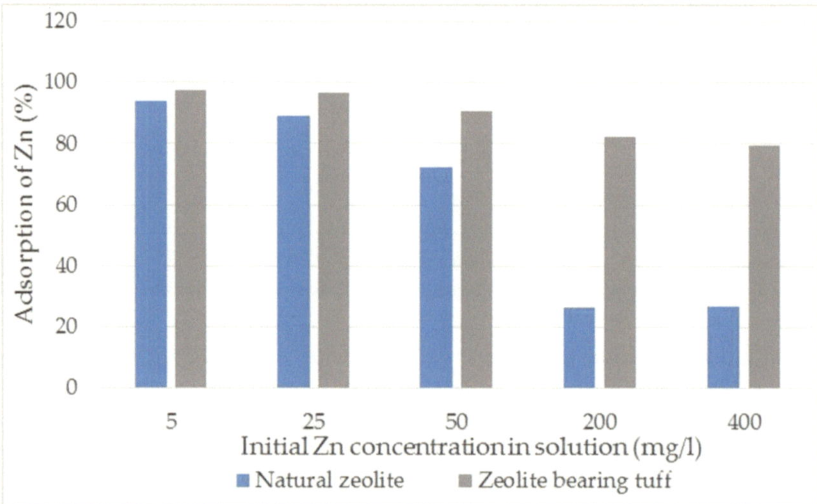

Figure 10. Adsorption of zinc depending on initial metal concentration.

http://dx.doi.org/10.5772/62347

By comparing the adsorbents used, natural zeolite and zeolite bearing tuff, for lead and zinc removal from solution, as shown in **Figures 9** and **10**, it can be concluded that zeolite bearing tuff is a more effective adsorbent, especially in the case of adsorption of zinc ions. In the case of adsorption of lead ions, which depends on adsorbents, significant changes are not found.

3.2. Effect of mass of adsorbent

The effect of mass of adsorbent was investigated by contacting 2, 5, and 10 g of adsorbent (natural zeolite or zeolite bearing tuff), at pH 3.5, with concentrations of single-component solutions of 5 mg/l Pb ions and 25 mg/l Zn ions.

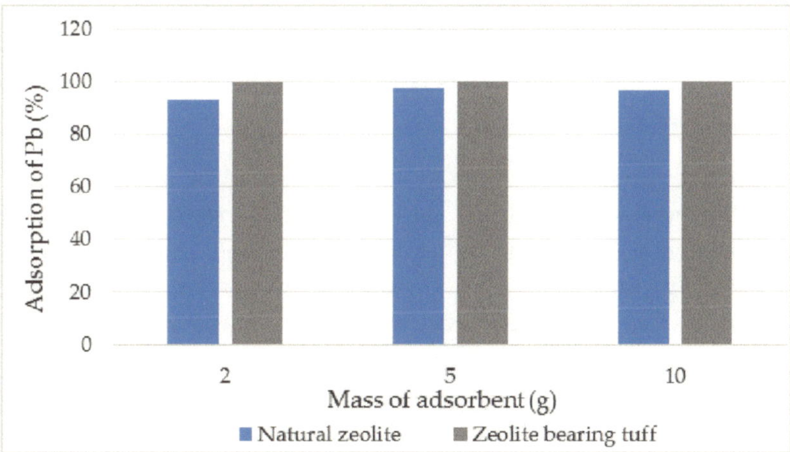

Figure 11. Adsorption of lead depending on the mass of adsorbents.

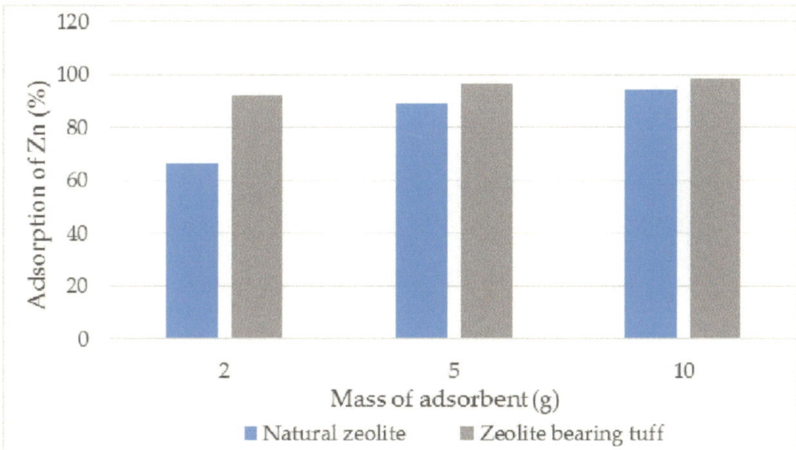

Figure 12. Adsorption of zinc depending on the mass of adsorbents.

Results from **Figures 11** and **12** show that an increase in the adsorbent mass resulted in an increase in the adsorption of the heavy metals. This was expected, because more adsorption sites are available per unit mass of adsorbent.

However, this is more apparently in the case of adsorption of zinc ions. Adsorption of lead ions increases with an increase in adsorbent mass, but less than the zinc ions.

By comparing the adsorbents used, and as shown in **Figures 11** and **12**, we come to the same conclusion, as that in the examination for the effect of initial concentration in solution, that the zeolite bearing tuff is the more effective adsorbent.

3.3. Effect of initial pH value of solution

The effect of initial pH of solution was investigated by contacting 5 g of adsorbent (natural zeolite or zeolite bearing tuff), at three various pH values (2.5, 3.5, and 4.5), with concentrations of single-component solutions of 5 mg/l Pb ions and 25 mg/l Zn ions.

According to Low et al., less adsorption at lower pH could be ascribed to hydrogen ions competing with the metal ions for adsorption sites [38]. This means that the adsorbent surface becomes more positively charged at higher H^+ concentration, thus reducing the attraction between the adsorbent and metal ions. Contrary to this, with increasing pH value, the adsorbent surface becomes more negatively charged, thus facilitating greater metal uptake [1].

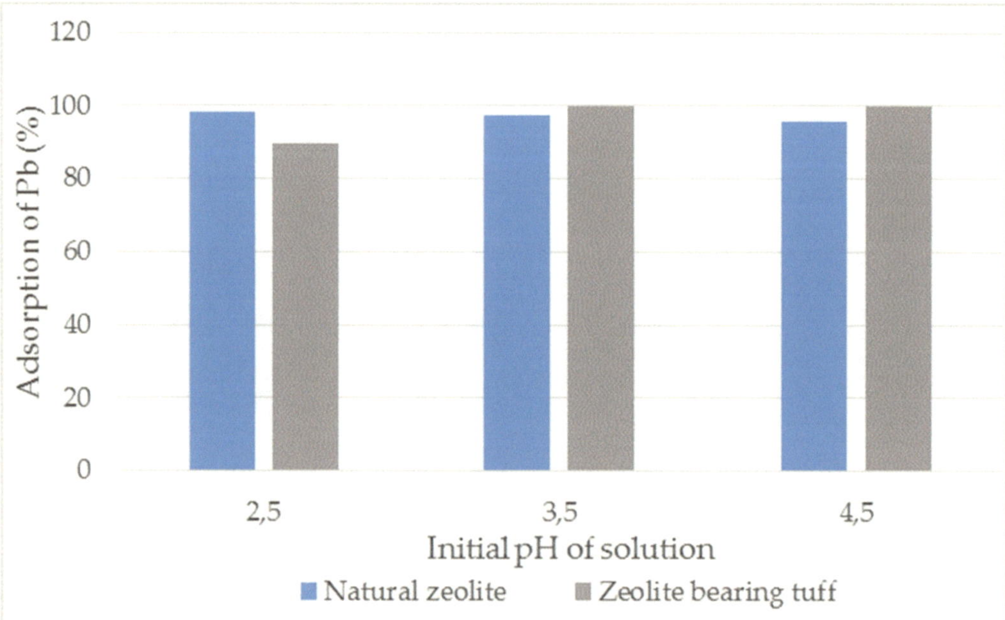

Figure 13. Adsorption of lead depending on initial pH value of solution.

http://dx.doi.org/10.5772/62347

As expected, as pH of a solution decreases, lead and zinc removal efficiency also decreases. This is because H^+ ions compete with lead and zinc cations for the same exchange sites [21], and the electrostatic repulsion between the lead and zinc cations in solution and the protonated zeolite surface increases as more H^+ ions are adsorbed [7].

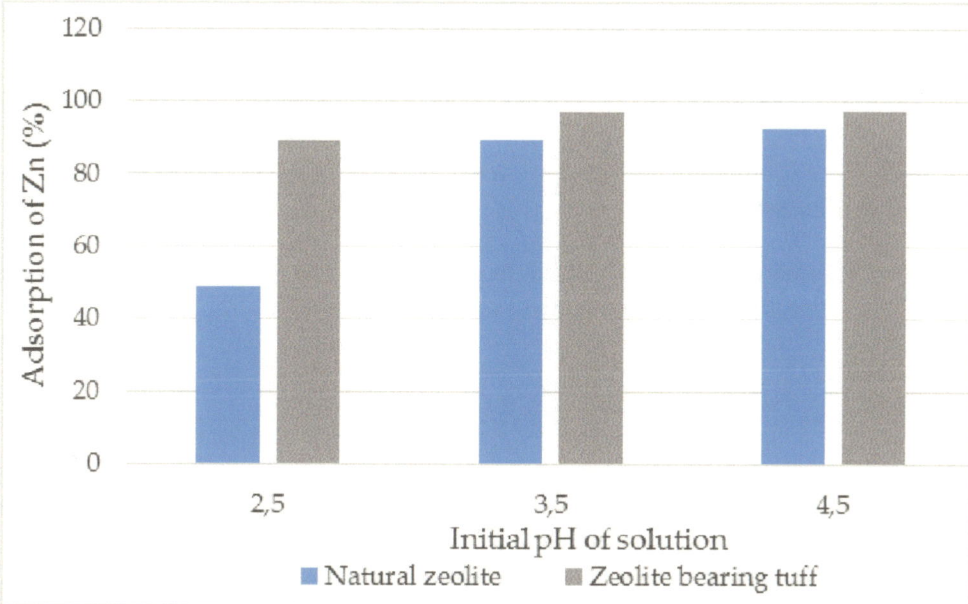

Figure 14. Adsorption of zinc depending on initial pH value of solution.

Figures 13 and **14** show how the adsorption capacity of natural zeolite and zeolite bearing tuff is affected by the solution's pH. Similar results on the efficiency of metal adsorption, dependent on the pH levels of solution, have been obtained by Moreno et al. [39], Alvarez-Ayuso et al. [21], and Motsi [10].

By comparing the adsorbents, we come to the same conclusion that zeolite bearing tuff has better adsorption than natural zeolite. The greater adsorption of heavy metals onto zeolite bearing tuff—stilbite compared to natural zeolite—clinoptilolite may be due to the higher porosity of zeolite bearing tuff.

4. Equilibrium studies

Equilibrium studies generally involve the determination of the adsorption capacity of a given material. This determination is important in accessing the potential of the material as an economic and commercially viable adsorber.

Adsorption will occur upon contacting an adsorbent with solutions containing an adsorbate. Until equilibrium is established between the adsorbate in solution and the adsorbed state, adsorption will continue. At equilibrium, a relationship exists between the concentration of the adsorbate in solution and the "concentration" of the adsorbate in the adsorbed state (i.e., the amount of adsorbed per unit mass of adsorbent) [40].

Determination of the maximum capacity of used adsorbents, natural zeolite and zeolite bearing tuff, for removal of lead and zinc ions from solution was the main objective of the equilibrium studies.

Experimental data were fitted to conventional adsorption mathematical models, namely the Freundlich and Langmuir models. These were used to predict the adsorption performance of natural zeolite and zeolite bearing tuff.

4.1. Langmuir's model

The Langmuir isotherm model [41], based on monolayer coverage of adsorbent surfaces by the adsorbate at specific homogeneous sites within the adsorbent, is represented as

$$q_{\mathring{a}} = \frac{q_m K_l C_e}{1 + K_l C_e} \tag{3}$$

where q_e, (mg/g) is the amount of solute adsorbed per unit mass of adsorbent at equilibrium; C_e, (mg/dm^3) is the residual adsorbate concentration in solution at equilibrium; q_m, (mg/g) is the amount of solute adsorbed per unit mass of adsorbent corresponding to complete coverage of available sites; K_l, (dm^3/mg) is the Langmuir adsorption coefficient, and this constant is related to the affinity between the adsorbent and solute, which is evaluated through linearization of Eq. (4):

$$\frac{1}{q_{\mathring{a}}} = \frac{1}{k_l q_m C_e} + \frac{1}{q_m} \tag{4}$$

The essential characteristics of Langmuir isotherm can be described by a dimensionless constant called equilibrium parameter, R_L, which is usually defined by

$$R_L = \frac{1}{(1 + K_l + C_0)} \tag{5}$$

where K_l is the Langmuir constant that indicates the nature of adsorption and C_0 is the highest initial metal concentration (mg/l). The value of R_L indicates the type of the adsorption isotherm to be either irreversible ($R_L = 0$), favorable ($0 < R_L < 1$), linear ($R_L = 1$), or unfavorable ($R_L > 1$).

4.2. Freundlich's model

The Freundlich isotherm model, based on monolayer adsorption on heterogeneous surfaces with a nonuniform distribution of adsorption heat, is represented as

$$q_{\mathring{a}} = k_f C_e^{1/n} \tag{6}$$

where k_f and n are empirical Freundlich constants that are dependent on experimental conditions. k_f (mg/g) is an indicator of adsorption capacity, while n (g/dm^3) is related to the adsorption intensity or binding strength. Their values were determined from the linear form of the Freundlich equation, given by

$$\log q_{\mathring{a}} = \log k_f + \frac{1}{n} \log C_e \tag{7}$$

where $1/n$ is the heterogeneity factor; values of $1/n << 1$ indicate heterogeneous adsorbents, while values closer to or even 1 indicate a material with relatively homogeneous binding sites [42]. Natural zeolite should be a heterogeneous adsorbent due to its porous nature. Alvarez-Ayuso et al. [21], Avila [43], and Gunay et al. [44] successfully used the Freundlich adsorption isotherm to model their results from equilibrium experiments.

4.3. Equilibrium experiments

Equilibrium experiments, the main objective of which was to determine the maximum capacity of adsorbents used, natural zeolite and zeolite bearing tuff, in removal of lead and zinc ions from solution, were done by contacting 5 g adsorbent at pH 3.5 and temperature of 20 ± 1°C with different concentrations of single-component solutions of 5, 25, 50, 200, and 400 mg/l of Pb^{2+} and Zn^{2+}.

Experimental data obtained from equilibrium experiments were fitted to the Langmuir and Freundlich adsorption isotherms. The values of the parameters for the two isotherms are presented in **Table 6**.

According to the Freundlich model, values of the heterogeneity factor, $1/n$, indicate that natural zeolite and zeolite bearing tuff are heterogeneous adsorbents, because the value of the heterogeneity factor is << 1. Based on the correlation coefficients (R^2), the adsorption isotherms can be better described by the Langmuir model. The applicability of the Langmuir isotherm suggests monolayer coverage of the Pb^{2+} and Zn^{2+} at the surface of the zeolites or that a similar exchangeable ion with equivalent concentration was exchanged by Pb^{2+} and Zn^{2+}. The R_L values reported in **Table 6** show that the behavior of Pb^{2+} and Zn^{2+} adsorption onto the adsorbents used was favorable ($0 < R_L < 1$).

		Lead		Zinc	
		Natural zeolite	**Zeolite bearing tuff**	**Natural zeolite**	**Zeolite bearing tuff**
Experimental	q_e(mg/g)	30.14	27.224	3.6392	25.408
	C_e(mg/l)	21.21	89.7	218.04	82.4
Langmuir	q_m(mg/g)	37.59	27.548	3.7272	29.0698
	k_l (l/mg)	0.1726	0.8875	0.0968	0.0464
	R_L	0.0143	0.0028	0.025	0.0511
	R^2	0.8853	0.9937	0.99	0.845
Freundlich	k_f (l/mg)	1.1628	9.3154	0.3934	1.6118
	1/n	0.6803	0.2667	0.4545	0.6128
	n (g/l)	1.4699	3.7495	2.2002	1.6319
	R^2	0.8346	0.7525	0.9111	0.9849

Table 6. Calculated equilibrium adsorption isotherm constants for the uptake of lead and zinc from solution by natural zeolite and zeolite bearing tuff.

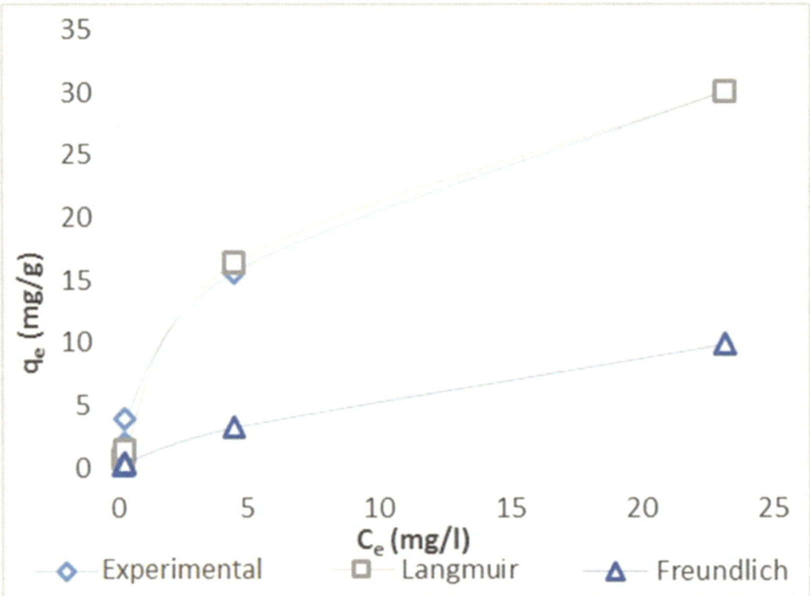

Figure 15. Equilibrium isotherms for Pb adsorption onto natural zeolite—clinoptilolite.

http://dx.doi.org/10.5772/62347

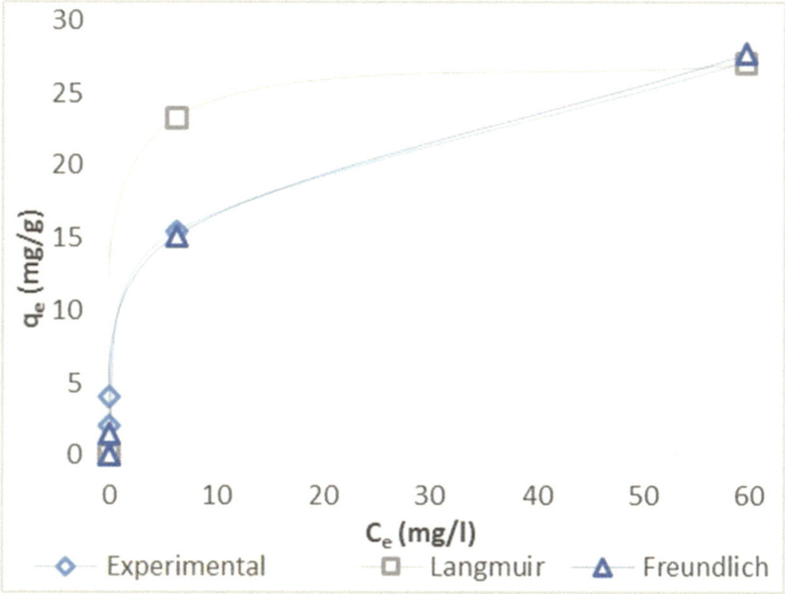

Figure 16. Equilibrium isotherms for Pb adsorption onto zeolite bearing tuff—stilbite.

Adsorption isotherms obtained according to the Langmuir and Freundlich models are compared with experimental data and are shown in **Figures 15** and **16** for the adsorption of lead and in **Figures 17** and **18** for the adsorption of zinc from the solutions of both the adsorbents used.

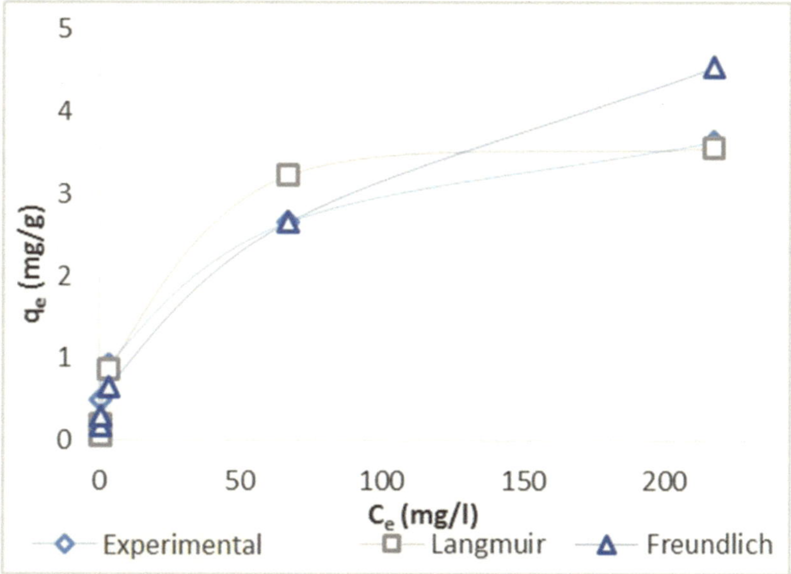

Figure 17. Equilibrium isotherms for Zn adsorption onto natural zeolite—clinoptilolite.

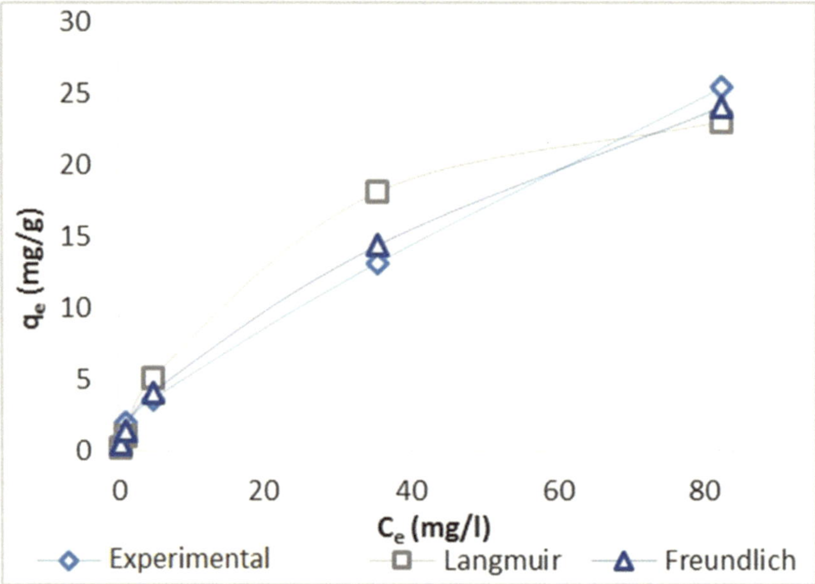

Figure 18. Equilibrium isotherms for Zn adsorption onto zeolite bearing tuff—stilbite.

The results shows that as the initial concentration of heavy metal cations increases, the amount of metal adsorbed per gram of adsorbent (q_e) increases. This is mainly due to the fact that at high metal concentrations, there is a higher solute concentration gradient, and this provides the necessary driving force for metal ions to displace exchangeable cations on the surface and

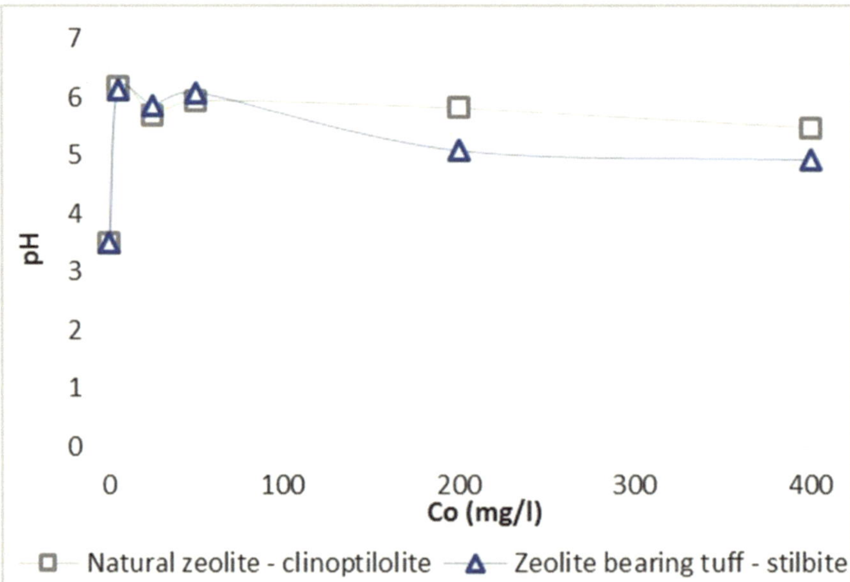

Figure 19. Equilibrium pH change with respect to various initial lead concentrations.

from the internal micropores of natural zeolite [10, 45]. However, this increasing trend is valid up to a point at which the maximum capacity of the adsorbent samples for the respective heavy metal cation is achieved, that is, its saturation point.

Figures 19 and **20** present the variation in the equilibrium pH values with respect to initial metal concentration in each equilibrium study.

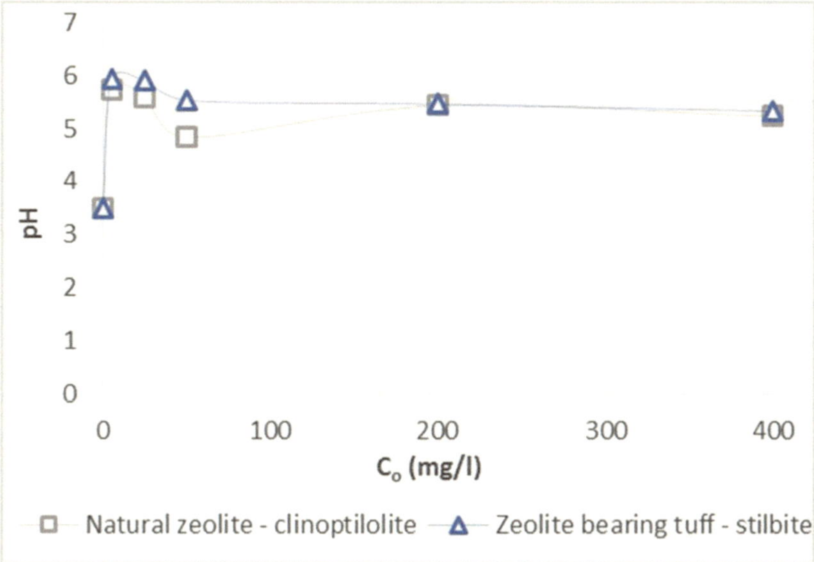

Figure 20. Equilibrium pH change with respect to various initial zinc concentrations.

By measuring the pH value before and after treatment, it can be concluded that the pH values at equilibrium are greater than initial pH values. The difference between equilibrium pH and initial pH exhibits a descending trend with increasing initial metal concentration. The adsorption of H^+ ions from solution will cause increase in pH value. As the initial concentration of metal increases, the concentration-driving force begins to favor the adsorption of metal ions in preference to H^+ ions, and thus the descending trend of the equilibrium pH at higher initial metal concentrations.

That natural zeolite has a buffering effect has been proven by many authors, including Erdem [8] and Motsi [10]. According to the results obtained from this research, it can be confirmed that not only natural zeolite—clinoptilolite but also zeolite bearing tuff—stilbite has a buffering effect.

5. Conclusion

The adsorption of heavy metal ions from diluted solution may occur efficiently onto porous materials. The studied natural raw materials which belong to the zeolite group are porous minerals with high adsorption capacity. Porosity of zeolite bearing tuff—stilbite (48.40%) is higher than the porosity of natural zeolite—clinoptilolite (20.20%), whereas the total cation exchange capacity of natural zeolite—clinoptilolite (1.8–2.2 mEq/g) is higher than that of zeolite bearing tuff—stilbite (0.69–1.07 mEq/g).

The adsorption of lead and zinc ions from diluted solution onto natural zeolite and zeolite bearing tuff occurs efficiently.

Regarding heavy metals, adsorption of lead ions is better than zinc ions. As expected, the results confirmed that ions with smaller hydration radius and hydration energy are better adsorbed on the studied porous raw materials. Adsorption of lead ions was found to be dependent on the pH of the solution, mass of adsorbents, and initial lead concentration in solution in the studied range. The adsorption of zinc ions is also efficient, but it is dependent on the conditions studied. It is most effective at lower concentration of zinc ions, higher mass of adsorbents, and higher pH value of the solution.

By comparing the studied adsorbents, it can be concluded that zeolite bearing tuff adsorbs better than natural zeolite. The greater adsorption of heavy metals onto zeolite bearing tuff—stilbite, when compared to natural zeolite—clinoptilolite, may be due to the higher porosity of zeolite bearing tuff.

Equilibrium studies show that adsorption of lead and zinc ions strongly depends on the initial concentration. As the initial concentration of heavy metal cations increases, the amount of metal adsorbed per gram of adsorbent (q_e) increases.

The Langmuir isotherms for the adsorption of lead and zinc from solution gave better fits of the experimental results than the Freundlich isotherms.

Author details

Mirjana Golomeova and Afrodita Zendelska*

*Address all correspondence to: afrodita.zendelska@ugd.edu.mk

Goce Delcev University, Stip, Macedonia

References

[1] Gamze Turan N., Mesci B., "Adsorption of Copper(II) and Zinc(II) ions by various agricultural by-products. Experimental studies and modelling," *Environment Protection Engineering*, vol. 31, no. 4, pp. 143–161, 2011.

[2] Kobya M., Demirbas E., Senturk E., Ince M., "Adsorption of heavy metal ions from aqueous solutions by activated carbon prepared from apricot stone," *Bioresource Technology,* vol. 96, no. 13, pp. 1518–1521, 2005.

[3] Ammann L., Cation exchange and adsorption on clays and clay minerals, Dissertation, Christian-Albrechts-Universität, Kiel, 2003.

[4] Sag Y., Aktay Y., "Mass transfer and equilibrium studies for the sorption of chromium ions onto chitin," *Process Biochemistry* vol. 36, pp. 157–173, 2000.

[5] Rahmani A. R., Mahvi A. H., Mesdaghinia A. R., Nasseri S., "Investigation of ammonia removal from polluted waters by Clinoptilolite zeolite," *International Journal of Environmental Science & Technology,* vol. 1, no. 2, pp. 125–133, 2004.

[6] Buasri A., Chaiyut N., Phattarasirichot K., Yongbut P., Nammueng L., "Use of natural clinoptilolite for the removal of Lead (II) from wastewater in batch experiment," *Chiang Mai Journal of Science,* vol. 35, no. 3, pp. 447–456, 2008.

[7] Cabrera C., Gabaldon C., Marzal P., "Sorption characteristics of heavy metal ions by a natural zeolite," *Journal of Chemical Technology and Biotechnology,* vol. 80, pp. 477–481, 2005.

[8] Erdem E., Karapinar N., Donat R., "The removal of heavy metal cations by natural zeolites," *Journal of Colloid and Interface Science,* vol. 280, no. 2, pp. 309–314, 2004.

[9] Taffarel S. R., Rubio J., "On the removal of Mn^{2+} ions by adsorption onto natural and activated Chilean zeolites," *Minerals Engineering,* vol. 22,, pp. 336–343, 2009.

[10] Motsi T., Remediation of acid mine drainage using natural zeolite, The University of Birmingham, 2010.

[11] Zendelska A., Golomeova M., Blažev K., Boev B., Krstev B., Golomeov B., A, Krstev, "Kinetic studies of manganese removal from aqueous solution by adsorption on natural zeolite," *Macedonian Journal of Chemistry and Chemical Engineering,* vol. 34, no. 1, pp. 1857–5625, 2015.

[12] Lobo J.A.C., Heterogeneous asymmetric epoxidation of cis-ethyl cinnamte over Jacobsen's catalyst immobilized in inorganic porous materials, RWTH Aachen University, Germany, 2005.

[13] Peskov M., "Zeolites," 2015. [Online]. Available: http://www.asdn.net/asdn/chemistry/zeolites.shtml.

[14] McCusker Lynne B., C. Baerlochek, "Zeolite structures," in *Introduction to Zeolite Science and Practice,* Elsevier, 2001.

[15] Payra P., Dutta P. K., "Zeolites: A primer," in *Handbook of Zeolite Science and Technology,* Marcel Dekker, 2003.

[16] Baerlocher C., Meier W. H., Olson D. H., *Atlas of Zeolite Framework Types*, 6th Edition, Elsevier, 2007.

[17] Szostak R., *Molecular Sieves: Principles of Synthesis and Identification*, 2nd Edition, Thomson Science, 1998.

[18] Cui H., Li L.Y., Grace J.R., "Exploration of remediation of acid rock drainage with clinoptilolite as sorbent in a slurry bubble column for both heavy metal capture and regeneration," *Water Research*, vol. 40, pp. 3359–3366, 2006.

[19] Yuan G., Seyama H., Soma M., Theng B.K.G., Tanaka A., "Adsorption of some heavy metals by natural zeolites: XPS and batch studies," *Journal of Environmental Science and Health*, vol. 34, pp. 625–648, 1999.

[20] Malliou E., Loizidou M., Spyrellis N., "Uptake of lead and cadmium by clinoptilolite," *Science of the Total Environment*, vol. 149, pp. 139–144, 1994.

[21] Alvarez-Ayuso E., Garcia-Sanchez A., Querol X., "Purification of metal electroplating waste waters using zeolites," *Water Research*, vol. 37, pp. 4855–4862, 2003.

[22] Sprynskyy M., Boguslaw B., Terzyk A.P., Namiesnik J., "Study of the selection mechanism of heavy metal (Pb^{2+}, Cu^{2+}, Ni^{2+} and Cd^{2+}) adsorption on clinoptilolite," *Journal of Colloid and Interface Science*, vol. 304, pp. 21–28, 2006.

[23] Leinonen H., Letho J., "Purification of metal finishing waste waters with zeolites and activated carbons," *Waste Management and Research*, vol. 19, pp. 45–57, 2001.

[24] Cincotti A., Mameli A., Locci M.A., Orru R., Cao G., "Heavy metal uptake by Sardinian natural zeolites: Experiment and modelling," *Industrial and Engineering Chemistry Research*, vol. 45, pp. 1074–1084, 2006 .

[25] Deer A., Howie R., Wise W.S., Zussman J., *Rock Forming Minerals* vol. 4B. *Framework Silicates: Silica Minerals, Feldspathoids and the Zeolites*, London: The Geological Society, 2004.

[26] "Clinoptilolite," 2015. [Online]. Available: http://www.iza-online.org/natural/Datasheets/Clinoptilolite/clinoptilolite.htm.

[27] Coombs D.S., Alberti A., Armbruster T., Artioli G., Colella C., Galli E., Grice J.D., Liebau F., Mandarino J.A., Minato H., Nickel E.H., Passaglia E., Peacor D.R., Quartieri S., Rinaldi R., Ross M., Sheppard R.A., Tillmanns E., Vezzalini, "Recommended nomenclature for zeolite minerals: Report of the Subcommittee on Zeolites of the International Mineralogical Association, Commission on New Minerals and Mineral Names," *Canadian Mineral*, vol. 35, pp. 1571–1606, 1997.

[28] "Stilbite," 2015. [Online]. Available: http://www.iza-online.org/natural/Datasheets/Stilbite/stilbite.htm.

[29] "Toxicological profile for lead," U.S. Department of Health and Human Services, Agency for Toxic Substances and Disease Registry, 2007.

[30] "Toxicology profile for zinc," U.S. Department of Health and Human Services, Agency for Toxic Substances and Disease Registry, 2005.

[31] Blazev K., Sijakova-Ivanova T., Panov Z., Zajkova-Paneva V., "Preliminary investigations into the mineralogy and potential uses of the stilbite rich tuffs from kratovo–zletovo volcanic area, Republic of Macedonia," *Comptes rendus de l'Academie Bulgare des Sciences,* vol. 65, no. 2, pp. 187–192, 2012.

[32] Bogdanovic G.D., Antic D.V., Begovic J., Stankovic V., "Adsorption of heavy metals from aqueous solution by natural zeolite," in *Proceedings of the XIV Balkan Mineral Processing Congress,* Tuzla, 2011.

[33] Nightingale E. R., "Phenomenological theory of ion solvation. Effective radii of hydrated ions," *Journal of Physical Chemistry,* vol. 63, no. 9, pp. 1381–1387, 1959.

[34] Mobasherpour I., Salahi E., Pazouki M., "Comparative of the removal of Pb^{2+}, Cd^{2+} and Ni+ by nano crystallite hydroxyapatite from aqueous solutions: Adsorption isotherm study," *Arabian Journal of Chemistry,* vol. 5, no. 4, pp. 439–446, 2012.

[35] Colella C., "Ion exchange equilibria in zeolite minerals," *Mineralium Deposita,* vol. 31, pp. 554–562, 1991.

[36] Connors K. A., *Chemical Kinetics: The Study of Reaction Rates in Solution,* USA: VCH Publishers, 1990.

[37] Barrer V. R. M., *Zeolites and Clay Minerals as Sorbents and Molecular Sieves,* London: Academic Press Inc., 1978.

[38] Low K.S., Lee C.K., Lee K.P., "Sorption of copper by dye treated oil-palm fibers," *Bioresource Technology,* vol. 44, no. 2, pp. 109–112, 1993.

[39] Moreno N., Querol X., Ayora C., "Utilization of zeolites synthesised from coal fly ash for the purification of acid mine waters," *Environmental Science and Technology,* vol. 35, pp. 3526–3534, 2001.

[40] Armenante P. M., "Adsorption," in *Industrial Waste Control: Physical and Chemical Treatment,* New Jersey Institute of Technology.

[41] Langmuir I., "The adsorption of gases on plane surfaces of glass, mica and platinum," *Journal of American Chemical Society,* vol. 40, pp. 1361–1403, 1918.

[42] Papageorgiou K. S., Katsaros K.F., Kouvelos P.E., Nolan W.J., LeDeit H., Kanellopoulos K.N., "Heavy metal sorption by calcium alginate beads from Laminaria digitata," *Journal of Hazardous Materials,* vol. B137, pp. 1765–1772, 2006.

[43] Avila M. A. S., *Experiment and Modelling of the Competititive Sorption and Transport of Chlorinated Ethenes in Porous Media,* Gottingen: Cuvillier Verlag, 2005.

[44] Gunay A., Arslankaya E., Tosun I., "Lead removal from aqueous solution by natural and pretreated clinoptilolite: Adsorption equilibrium and kinetics," *Journal of Hazardous Materials,* vol. 146, pp. 362–371, 2007.

[45] Çagin V., Use of clinoptilolite for copper and nickel removal from aqueous solutions, The Graduate School of Natural and Applied Sciences of Middle East Technical University, 2006.

[46] Zendelska A., Golomeova M., Blažev K., Krstev B., Golomeov B., Krstev A., "Kinetic studies of zinc ions removal from aqueous solution by adsorption on natural zeolite," *International Journal of Science, Environment and Technology,* vol. 3, no. 4, pp. 1303–1318, 2014.

[47] Zendelska A., Golomeova M., Blažev K., Krstev B., Golomeov B., A, Krstev, "Equilibrium studies of zinc ions removal from aqueous solution by adsorption on natural zeolite," *Journal of Materials Science and Engineering A,* vol. 4, no. 7, pp. 202–208, 2014.

Chapter 3

Synthesis and Characterization of CMK Porous Carbons Modified with Metals Applied to Hydrogen Uptake and Storage

Marcos B. Gómez Costa, Juliana M. Juárez and
Oscar A. Anunziata

Additional information is available at the end of the chapter

http://dx.doi.org/10.5772/63710

Abstract

In this chapter, we have shown that hopeful hydrogen storage material can be obtained by ordered mesoporous carbons (carbons mesostructured from Korea, CMK-1 and CMK-3) and modified with metal/cations species. The pristine CMK-1 and CMK-3 were synthesized by replication using MCM-48 and SBA-15 as hard templates and sucrose as a carbon source. Incorporation of metal species was carried out by wetness impregnation. The mesoporous materials modified were characterized by X-ray diffraction (XRD), X-ray photoelectron spectra (XPS), RAMAN, transmission electron microscopy (TEM), and adsorption/desorption N_2 isotherms. Carbon modified with metal/cations shows a better capacity for hydrogen uptake than that of the mesoporous carbons. The evolution of high-pressure hydrogen adsorption measured at 77 K shows that composites can significantly enhance hydrogen adsorption capacity and hydrogen storage performance of carbon materials, proving to be prospective candidates for application in hydrogen storage. The improved activity and the larger performance of composite materials are attributed to improved dispersion of uniform metal/cations nanoparticles as well as to efficient use of the support, which may originate a high-surface area and pore volume, allowing a large dispersion of clusters.

Keywords: mesoporous materials, nanocomposites, modified CMK, characterization, hydrogen storage

1. Introduction

Hydrogen is considered as a clean energy carrier and an alternative fuel source for many applications. Some keys to use hydrogen, as a fuel, are its safe storage, low cost, lightweight, and it reversibility, with a simple adsorption-desorption kinetics.

The great challenge posed by the use of hydrogen as an energy source today is primarily focused on the need for safe and stable storage. Its capacity as an energy carrier is demonstrated by its increased energy per mass, cleanliness and ease of production from renewable sources, as opposed to fossil fuels [1]. Thus, a viable solution for both storage and transport problems is required, since alternatives such as hydrogen liquefaction or compression present difficulties due to its low storage density and the high costs involved [1].

Many activated and templated carbons with high-surface area exhibit potential for storing molecular hydrogen [1–4]. Their small weight, low cost, and wealth of natural precursors make nanostructured carbons with high-surface area likely candidates for physisorption. The literature reports that activated carbons show relatively high hydrogen storage capacity at 77 K [2, 3]; however, pore size distribution is in most cases wide, and more than 50% of the total porous volume is given by macropores, whose contribution to hydrogen uptake is more limited.

Carbons mesostructured from Korea (CMK) comprise one such family of ordered mesoporous carbons (OMC) [5]. Produced inside the channels of mesostructured silicates or aluminosilicates, CMK have specific surface areas from 1000 to 2000 m^2/g and pore volumes from 0.5 to over 1 ml/g. These materials are promising for hydrogen storage applications. The template method was used to obtain the carbons, involving first introduction of the appropriate carbon precursors into the ordered template pores and then carbonization and removal of the template. Such carbon materials are characterized by large surface areas and high pore volumes, which make effective H_2 physisorption feasible.

However, a common limitation is found in weak van der Waals interaction between molecular hydrogen and host material. Hence, to enhance hydrogen binding, low amounts of transition metals or cations are added, improving hydrogen sorption [6–9].

Thus, CMK-1 and CMK-3 were chosen as ideal support material for hydrogen storage due to their large surface area, high chemical stability, uniform pore diameter, accessible porosity, and three-dimensional conducting network [10,11].

The scattering of metal particles throughout the porosity of active carbons greatly enhances their storage abilities. Numerous studies devoted to this feature have given ample evidence of hydrogen spillover. H_2 dissociation is supposedly triggered by the catalyst, with atomic hydrogen diffusing deeper into the microporous network and even in between graphitic layers [12].

Nanometric clusters of Pt, Zn, Ni, V, and TiO_2 dispersed in CMK-1 or CMK-3 framework have been reported as promising hydrogen storage nanomaterials. The adsorption in these materials was fully reversible and higher than that in carbon hosts [12,13].

Nanocluster activity improved hydrogen adsorption molecules, producing substantial changes in hydrogen sorption properties.

2. Material and methods

2.1. Templates preparation: MCM-48 and SBA-15

The mesoporous material MCM-48 was prepared as the template for the CMK-1. The mesoporous silicate was synthesized by hydrolysis of tetraethylorthosilicate (TEOS, 98%, Sigma–Aldrich) at room temperature, in an aqueous solution, using cetyltrimethylammonium bromide (CTAB, Sigma–Aldrich) as a surfactant. The procedure designed is described as follows: Sodium hydroxide (NaOH, Sigma–Aldrich) and CTAB were dissolved in deionized water, and then tetraethyl orthosilicate (TEOS) was added. The molar composition of the gel was 1 M TEOS: 0.48 M NaOH: 0.48 M CTAB: 55 M H_2O. After stirring for about 1 h, the solution was transferred into a Teflon bottle and there it was heated up to a temperature of 373 K. After 72 h, the mixture was kept at RT and a solution of HCl was used to turn the pH to 7. Afterwards, the solution was kept at 373 K for 72 h. In order to wash and dry the product, the solution was filtered [14]. The template was extracted by thermal treatment performed up to 373 K, using N_2 (20 mL/min) and afterwards a calcination at 723 K was implemented, raising the temperature 3 K/min [15].

Ordered mesoporous silica SBA-15 was prepared using the triblock copolymer, poly(ethylene glycol)-block-poly(propylene glycol)-block-poly (ethylene glycol), (EO20PO70EO20, P123-Sigma–Aldrich), as surfactant, and tetraethyl orthosilicate (TEOS, Sigma–Aldrich, reagent grade 98%) as the source of silica. The technique planned was: A solution was prepared with P123 (20 g) in 400 ml of HCl 1 M at 323 K. Then, TEOS (40 g) was drop under stirring during 24 h at the same temperature. The subsequent white combination was moved to a propylene flask and matured by 72 h at 373 K. The material was filtered out and then washed with deionized water until pH ~ 6. Its molar composition was 1 Si:0.018 $EO_{20}PO_{70}EO_{20}$:2.08 HCl:112 H_2O. The template was removed by immersing the material in ethanol reflux for 6 h and then calcining it at 823 K in air for 6 h. The product was finally filtered, washed, and dried in air at 363 K [16].

2.2. Synthesis of the replicas CMK-1 and CMK-3

Mesoporous carbons CMK-1 and CMK-3 were prepared via a two step impregnation of the mesopores of MCM-48 and SBA-15 with a solution of sucrose using an incipient wetness method. Briefly, 1.0 g of the as-prepared silica material MCM-48 for CMK-1 and SBA-15 for CMK-3 was soaked with 1.1 g of sucrose and 0.14 g of H_2SO_4 in 5.0 g of deionized water solution. The mixture was then dried at 373 K for 6 h and then at 433 K for 6 h. The silica sample, which now contained partially polymerized and carbonized sucrose, had 0.65 g of sucrose, 90 mg of H_2SO_4, and 5.0 g of deionized water added to it, and was treated again at 373 K and 433 K. To complete the carbonization, the sucrose-silica composite was heated at 1173 K for 4 h under nitrogen flow (20 mL/min).

The silica template was dissolved with 5 wt% hydrofluoric acid at room temperature. This step was done two times in order to dissolve entirely the template. The template-free carbon product thus obtained was filtered, washed with deionized water and ethanol, and dried [12, 13].

2.3. Synthesis of Zn-CMK-1

Zn-CMK-1 was prepared using as a source of zinc Zn nitrate (Sigma–Aldrich 98%, hexahydrate), dissolved in ethanol, and then mixed with a portion of CMK-1 at room temperature. The suspension was left static for 24 h, before being dried at 373 K for a period of 18 h, and subsequently desorbed from 298 K to 473 K in an inert atmosphere (N_2), at a rate of 10 K/min. It was held at this latter temperature for 5 h. During the whole process, the nitrogen flow was 20 ml/min. To obtain the reduced sample of Zn-CMK-1, it was placed under H_2 flow at 773 K for more than 6 h with a ramp rate 10 K/min. The sample was denoted as Zn-CMK-1. ICP analysis showed that the pristine sample of Zn-CMK-1 contained 5.5 wt% of Zn before reduction and 5 wt% afterwards [13].

2.4. Synthesis of Ni-CMK-1

Ni-CMK-1 was obtained by impregnating CMK-1 using $NiCl_2.6H_2O$ (Sigma–Aldrich 98%) dissolved in ethanol, which was stirred vigorously for 20 min. Then, the sample was dried under vacuum at 308 K, to achieve complete dryness by evaporation of the solvent. The powder obtained was heated to 353 K to dry and then placed in a tubular furnace at 773 K under controlled flow of H_2 for reduction during 6 h. It was determined by ICP that the reduced sample contained 5.5 wt% Ni [13].

2.5. Synthesis of Pt-CMK-3

The sample was prepared by wetness impregnation using chloroplatinic acid ($H_2PtCl_6 \cdot H_2O$-Sigma–Aldrich) as Pt source. Chloroplatinic acid in ethanol was added to the carbon solution at room temperature. Excess of ethanol was removed by placing the solution in a rotary evaporator at around 333 K and 60 revolutions per minute. The sample was then dehydrated at 373 K during 18 h. The resultant solid was heated under nitrogen flow from 298 K to 473 K (ramp 10 K/min) and maintained at 473 K for 5 h. Afterwards, it was heated up to 743 K (slope 10 K/min), where it was kept for another 5 h. The nitrogen flow was kept constant at 20 ml/min. The sample was also subjected to H_2 flow at 773 K for 6 h with a ramp rate of 10 K/min. The sample was denoted as Pt-CMK-3. 1% wt of Pt in the final Pt-CMK-3 material was obtained [12].

2.6. Synthesis of TiO_2-CMK-3

TiO_2 nanoparticles were incorporated into ordered mesoporous carbon CMK-3 using wetness impregnation and tetrabutyl orthotitanate (TBOT) as a titanium source. The metal precursor

http://dx.doi.org/10.5772/63710

(TBOT) was dissolved in 10 mL of ethanol with vigorous stirring, to reach a nominal content of 5 wt% of Ti in the final solid. CMK-3 was incorporated in the solution, and both were placed in a rotary evaporator to remove excess of ethanol at about 313 K and 60 rpm. The powder was then dried at 373 K overnight. Finally, the resulting material was heated in a dynamic inert atmosphere (nitrogen flow of 20 mL/min) from 298 to 473 K with a slope of 4 K/min; temperature was then increased to 743 K with a slope of 10 K/min and kept at this temperature during 5 h.

2.7. Synthesis of V-CMK-3

Vanadium nanoparticles were incorporated into ordered mesoporous carbon CMK-3 using wetness impregnation and VCl_3 as source of Vanadium. The metal precursor (VCl_3) was dissolved in 20 mL of ethanol with vigorous stirring, to have a nominal content of 1 wt% of V in the final solid. The solution was placed in a rotary evaporator to remove excess of ethanol at about 323 K and 50 rpm. The obtained powder was then dried at 373 K overnight. The resulting material was heated in a dynamic inert atmosphere (nitrogen flow of 20 mL/min) from 298 to 473 K with a slope of 4 K/min, and then, the temperature was increased to 743 K with a slope of 10 K/min and kept at this temperature during 5 h. A reduction procedure in H_2 flow (20 mL/min) was performed to the sample at 1023 K using the same procedure described earlier.

2.8. Characterization of the samples

ICP (ICP-AES; VISTA-MPX) was used to determine the elemental composition, working at a frequency of 1.5 kW under airflow 12 L/min. In order to determine textural properties, such as surface area and pore volume, N_2 adsorption/desorption isotherms at 77 K were measured on ASAP 2020. To realize studies of the adsorption isotherms, the samples were purged at 673 K, and size pore distribution assessed by means of Barrett, Joyner, and Halenda (BJH) algorithm. The hydrogen chemisorption characterization was realized in the Micromeritics Chemisorb 2720 apparatus, equipped with a detector TCD, at RT. The cleaned samples were titrated with H_2 pulses at 295 K under flow of N_2 up to a persistent output signal of TCD, which indicates saturation.

X-ray diffraction profiles of samples were recorded with a X'PertPro PANalytical diffractometer equipped with a Cu Kα radiation (λ = 0.154 nm, X-ray generator current and voltage set at 40 mA and 45 kV). The samples were crushed previously and placed on an aluminum sample holder. The diffraction patterns were recorded in steps over a range of Bragg angles (2θ) between 0.5° and 5° (and for wide angle between 20 and 70°), at a scanning rate of 0.02 per step and an accumulation time of 20 s. Diffractograms were analyzed with the X'Pert High Score Plus software. X-ray photoelectron spectra (XPS) were obtained on a Microtech Multilb 3000 spectrometer, furnished with a hemispherical electron analyzer and MgKα (hν = 1253.6 eV) photon source. An estimated error of ±0.1 eV can be obtained for all ranges. Peak intensity was determined from each peak areas after background removal and spectrum suitable by a combination of Gaussian/Lorentzian functions. Surface relation ratios (50–100 A of Deep) were extracted as a function of the intensities of the signals corrected by the sensitivity factors with an estimated precision of ±7%.

Raman spectrum was acquired from an InVia Reflex Raman microscope and spectrometer using a 532 nm diode laser excitation. The transmission electron microscopy (TEM) micrographs were recorded in a Philips EM 301, with accelerating voltage of 200 kV (point resolution of 0.17 nm).

To determine hydrogen storage isotherms, we employed an ASAP 2050 apparatus properly calibrated, at 77 K and at low and high pressures (up to 10 bar). Previous to all the adsorption experiments, the samples were degassed at 523 K during 8 h under vacuum conditions, to prevent the probable decomposition of the metal nanoparticles at higher temperature. Afterwards, the samples were cooled at RT. The adsorption experiments were carried out with high-purity hydrogen gas (99.9999%), and in a pressure range from 0 to 10 bar.

3. X-ray diffraction studies

Figure 1 shows typical XRD patterns for the MCM-48 silica template and the carbon CMK-1 produced with the aforementioned procedures. The low-angle XRD pattern of MCM-48 points to a high degree of structural order for the cubic crystallographic space group Ia3d (**Table 1**). However, a different pattern relative to MCM-48 can be observed for the carbon CMK-1, obtained by removing the silica wall after carbonization. As opposed to CMK-2, 3, and 4, the structure of the CMK-1 obtained through the use of silicate MCM-48 templates was not found to be an exact negative replica of the template. This was due to a transformation of the mesostructure after the template wall had dissolved. This structural transformation was mainly evidenced by the appearance of a strong low-angle diffraction peak (h k l [1 1 0] at 2θ = 1.67°) on the X-ray diffraction (XRD) patterns of CMK-1 (see **Figure 1**), which was not reliable with the symmetry of MCM-48 [12,17].

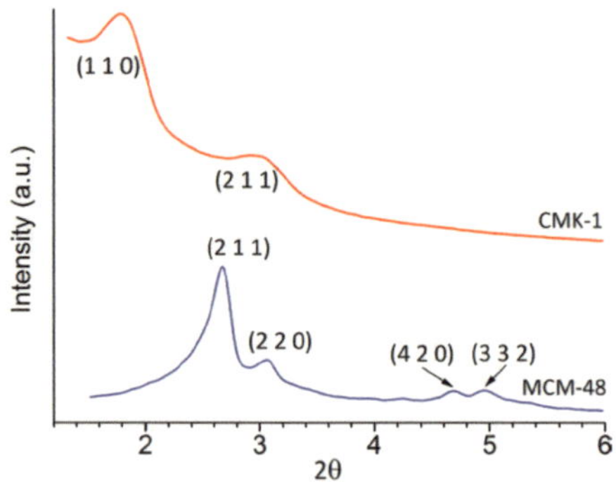

Figure 1. Small-angle XRD patterns of MCM-48 and CMK-1.

As shown by the low-angle powder XRD pattern, a structural transformation of the interlaced carbon framework occurs after the silica wall has dissolved. A displacement of the sub-frameworks with respect to each other may lead to the formation of contacts. A distortion of the frameworks may also take place as a consequence of further transformations. The principal indicator of structural transformation of the material is the strong low-angle diffraction peak appearing in the XRD powder pattern, which is inconsistent with the space group Ia3d of the MCM-48 template. This reflection is indexed as [1 1 0], and the explanation of its emergence may lie in a decrease in the material symmetry. The complexity of the framework structure of CMK-1 is such that it fails to be described by simple terms such as "pore," "wall," or "diameter," which are commonly used in descriptions of the texture of mesoporous materials and molecular sieves. However, an estimation of its geometric characteristics might be in place. It is possible to determine the diameter of the roughly cylindrical framework segments, which coincide with the Wyckoff symmetry positions 12c and 12d of the I_41_{32} space group [18].

Sample	MCM-48 (h k l)			CMK-1 (h k l)			a_0 (nm)
	2 1 1	2 2 0	4 2 0	3 3 2	1 0	2 1 1	
	d (nm)			d (nm)			
Si-MCM-48	3.71	3.21	2.02	1.93	–	–	9.1
CMK-1	–	–	–	–	5.85	3.35	8.2
Zn-CMK-1	–	–	–	–	5.59	–	7.9
Ni-CMK-1	–	–	–	–	5.78	3.32	8.2

Table 1. XRD parameters of samples.

Figure 2 shows the low-angle X-ray diffraction patterns of CMK-1, Ni-CMK-1, and Zn-CMK-1 samples. Retention of the overall pore structure after the addition of the different metals is indicated by the low-angle diffraction peaks (**Table 1**). However, the peak intensity of the samples is noticeably lower than that of the parent CMK-1. This intensity loss may be due to the introduction of scattering material into the pores, leading to an increased phase cancellation

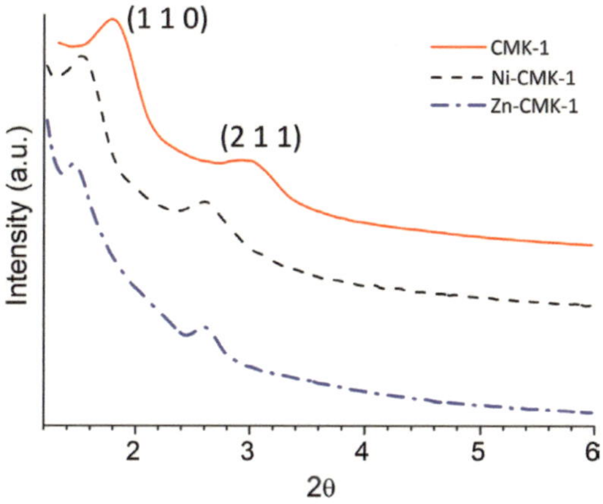

Figure 2. Small-angle XRD patterns of CMK-1, Ni-CMK-1, and Zn-CMK-1.

between scattering from the walls and the pore regions [19]. In the case of the Zn-CMK-1 sample, the absence of XRD reflections can be ascribed to a very high degree of pore filling.

Figure 3 shows a typical XRD pattern for the silica template SBA-15 and the carbon CMK-3 yielded by the procedures described. The low-angle XRD pattern of SBA-15 points to a high degree of structural order for the hexagonal P_6mm crystallographic space group (**Table 2**). A similar pattern relative to SBA-15 can be observed for the carbon CMK-3 obtained by removing the silica wall after carbonization. The structure of the CMK-3 obtained by the use of a silicate SBA-15 template was found to be an exact negative replica of the template. The structural transformation was mainly evidenced by the appearance of CMK-3 on the X-ray diffraction (XRD) patterns, which was consistent with the symmetry of SBA-15 [19].

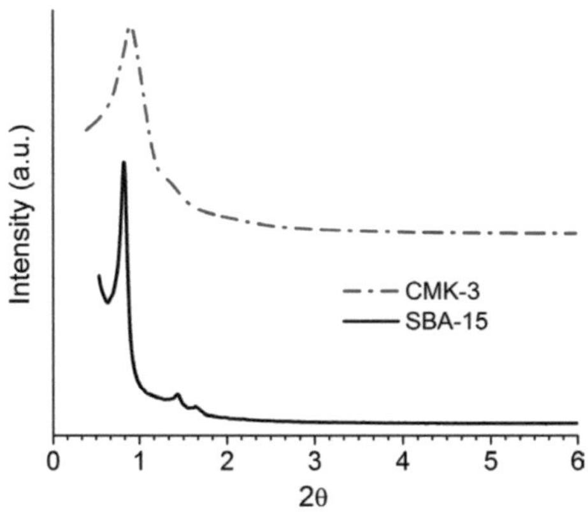

Figure 3. Small-angle XRD patterns of SBA-15 and CMK-3.

Sample	SBA-15 [h k l]			CMK-3 [h k l]			a_0 (nm)
	[1 0 0]	[1 1 0]	[2 0 0]	[1 0 0]	[1 1 0]	[2 0 0]	
	d (nm)			d (nm)			
SBA-15	8.6	5.0	4.3	–	–	–	10.0
CMK-3	–	–	–	10.5	6.1	5.2	12.1
Pt-CMK-3	–	–	–	8.4	4.9	4.2	9.70
TiO_2–CMK-3	–	–	–	7.1	4.4	3.8	8.60
V-CMK-3	–	–	–	7.2	4.2	3.9	8.50

Table 2. XRD parameters of samples.

Figure 4 shows the low-angle X-ray diffraction patterns of CMK-3 and of the modified samples. The overall pore structure is retained after the addition of Pt, TiO_2, and V, which is indicated by the low-angle diffraction peaks (**Table 2**). However, the signal intensity corresponding to the [1 1 0] diffraction plane of the samples shows a slow decrease in relation to the parent

CMK-3. The introduction of scattering material into the pores generally leads to an increased phase cancellation between scattering from the wall and the pore regions. The intensity loss of CMK-3 typical Bragg reflections is caused by the introduction of scattering material (Pt, TiO_2, and V) into the pores [19].

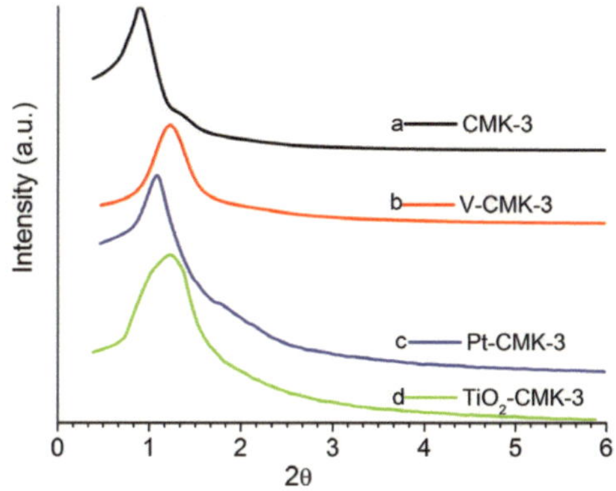

Figure 4. Small-angle XRD patterns of CMK-3 (a), V-CMK-3 (b), Pt-CMK-3 (c), and TiO_2–CMK-3 (d).

Figure 5 shows the wide-angle diffraction region of the Zn-modified sample (Zn-CMK-1). XRD pattern at the wide-angle range (10–60°) of the CMK-1 host (see the inset of **Figure 5**) exhibits two broad diffraction peaks which can be indexed as [0 0 2] and [1 0 0] diffraction for typical graphite carbons [20].

The pattern of Zn-CMK-1 shows ZnO and metallic Zn signals [21,22], probably due to the reduction process not being complete. The fact that this pattern exhibits no prominent reflections is an indicator that no crystalline bulk material was formed outside the pore system. The Zn/ZnO clusters are confined for the most part in the nanostructured carbon [23], showing nanometric size and high dispersion which can be seen in **Table 1** and TEM studies.

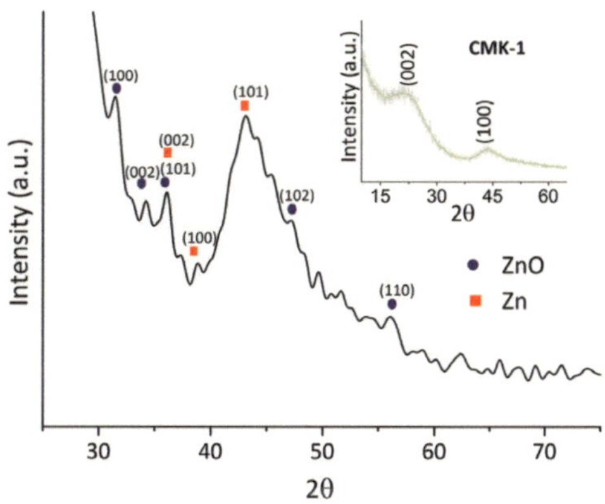

Figure 5. Wide-angle XRD patterns of Zn-CMK-1 and CMK-1 (inset).

The XRD pattern of Ni-CMK-1 (**Figure 5**) shows intensive peaks at 2θ = 44.5, 51.8, and 76.3°, corresponding to metallic Ni diffraction planes [1 1 1], [2 0 0], and [2 2 0], [45]. These are characteristic patterns of an fcc lattice consisting of nickel nanoparticles. NiO species may be present, since the NiO signals were very close to metallic Ni signals and would overlap [24] in the pattern, as TEM-EDS and XPS studies will show later. In agreement with XRD studies, the characteristic structure of CMK-1 is maintained after the metal is within the host, whereas the nanomaterial area becomes significantly smaller (**Table 1**).

Figure 6 displays the diffraction region of 20–70° 2Θ of samples. Typical XRD pattern of mesoporous materials was exhibited by SBA-15 sample. Two broad diffraction peaks were found in CMK-3 pattern which can be ascribed to [0 0 2] and [1 0 0] diffraction indexes for a characteristic graphite carbons [20].

The final sample of Pt-CMK-3 pattern displays metallic platinum peaks [25], pointing to a well-developed reduction process. In the case of Pt clusters, the absence of prominent reflections indicates that no crystalline bulk material formation had occurred outside the pore system [23], with nanometric size and high dispersion (very broad XRD signal of Pt, **Figure 6**) as shown in **Table 2**. The reason for this may be found in a relatively low scattering contrast between the pores and the walls of the mesoporous materials caused by the formation of reduced platinum which indicates fine dispersion on CMK-3 with a narrow size distribution [24].

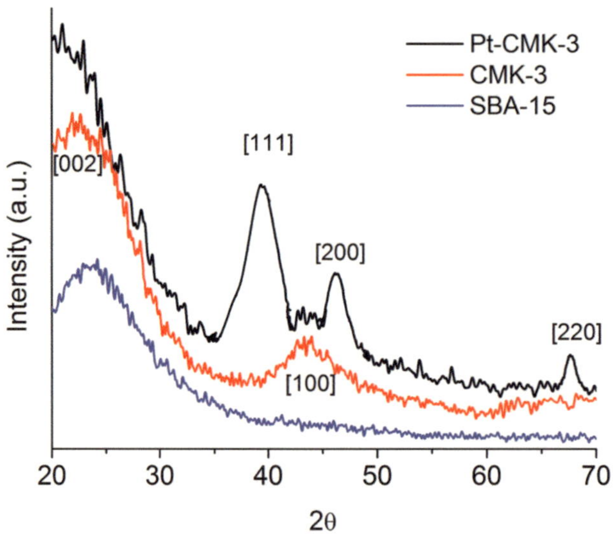

Figure 6. Wide-angle XRD patterns of SBA-15, CMK-3, and Pt-CMK-3.

Figure 7 shows the wide-angle X-ray diffraction patterns for CMK-3 and TiO_2-CMK-3. In both cases, two broad diffraction peaks are distinguished, which can be indexed as [0 0 2] and [1 0 0] diffraction for typical graphite carbons [20].

The pattern of TiO_2-CMK-3 shows characteristic signals of TiO_2 in anatase phase [26], indicating a particularly good synthesis process. The absence of prominent reflections in TiO_2 clusters indicates that no crystalline bulk material has been formed [23], with nanometric size and high dispersion (very broad XRD TiO_2 signals, **Figure 7**). This can be ascribed to a relatively low scattering contrast between the pores and walls of mesoporous materials, due to the formation of anatase nanoclusters that depict fine dispersion on CMK-3 with a narrow size distribution [24].

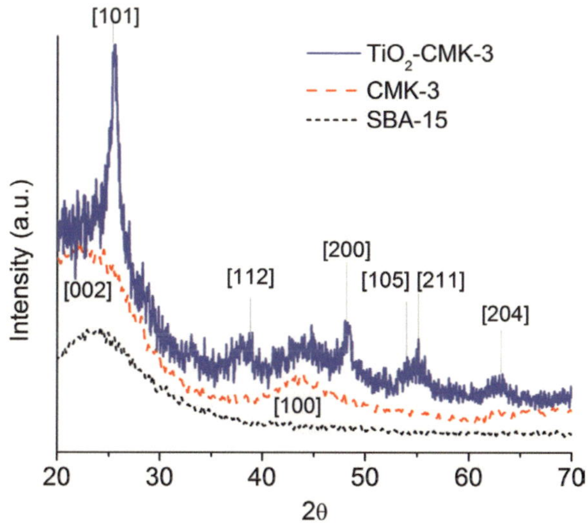

Figure 7. Wide-angle XRD patterns of SBA-15, CMK-3, and TiO_2–CMK-3.

Figure 8 shows the wide-angle X-ray diffraction patterns for CMK-3 and V-CMK-3. In both cases, two broad diffraction peaks are distinguished which can be indexed as [0 0 2] and [1 0 0] diffraction for typical graphite carbons [20].

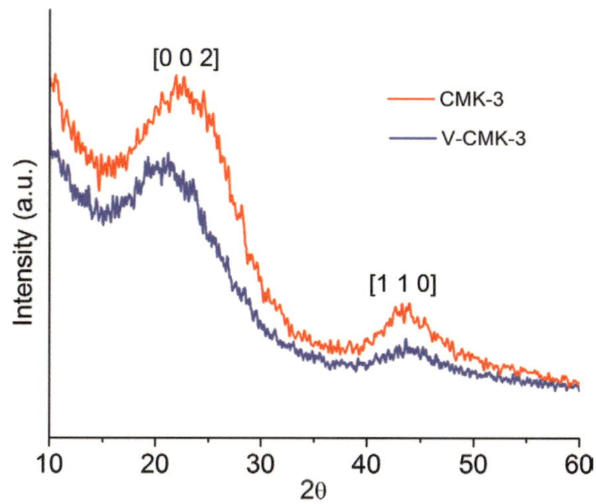

Figure 8. Wide-angle XRD patterns of CMK-3 and V-CMK-3.

In the wide-angle region, practically no reflections typical of vanadium have been found. The absence of these prominent reflections indicates that no crystalline bulk materials have been formed outside the pore system and it is an evidence that the clusters have nanometric size and high dispersion [23].

3.1. Cluster size estimation using Scherrer's formula

Analysis of the XRD patterns makes it possible to determine the size of crystals of metallic particles in the samples and to evaluate the effect on metal dispersion. The average cluster size was thus estimated based on the width of the diffraction peaks corresponding to different [h k l] Miller indices using Scherrer's formula (Eq. (1)) [27].

$$d = \frac{K\lambda}{\beta \cos\theta} \tag{1}$$

where d is the average size of the crystals or domains (the crystal mean size defined as the cube root of the material volume). K is a shape factor of the nanocluster, dimensionless, generally has a typical value of 0.89 or 0.83 (if the particles are approximately spherical or not), λ is the wavelength of the X-rays (1.54 Å) used for performing the analysis, β is the full width at half-maximum (FWHM, in radians) of the XRD signal, and θ is the Bragg angle (in radians) [28].

In order to apply Scherrer's formula, [1 1 1], [2 0 0], and [2 2 0] refractions plains for Pt-CMK-3 were employed. The average diameter of the metal particles, as obtained by Scherrer's formula, was ~1.76 nm. In the case of TiO_2-CMK-3, the average diameters of anatase particles obtained by Scherrer's formula were ~5 nm, employing [1 0 1], [1 1 2], and [2 0 0] refractions plains.

In the case of the sample CMK-1 modified with Ni, the planes [101], [200], and [220] were used. For the Zn-CMK-1, [100] and [101] refraction planes for Zn^0 and for ZnO the planes [002] and [110] were used. Using Scherrer's formula, the average size of metal nanoclusters for the sample modified with Ni were 13 nm, and 2.3 nm and 2.6 nm for the Zn^0 and the ZnO particles.

4. Nitrogen adsorption/desorption isotherm analysis

Figure 9 shows the N_2 adsorption/desorption isotherms for MCM-48, CMK-1, Zn-CMK-1, and Ni-CMK-1 samples, while **Table 3** shows the textural properties obtained from nitrogen physisorption analysis. The nitrogen adsorption–desorption is other for MCM-48 has a typical type IV curve (according to the IUPAC classification), which is a sure indicator of the mesoporous nature of the material. The isotherm for CMK-1 and Zn-CMK-1 and Ni-CMK-1 shows hysteresis loops at a relative pressure range of 0.4–0.6, which can be caused by capillary condensation-evaporation from the mesopores. Zn-CMK-1and Ni-CMK-1 clearly reveal a reduced specific surface area (as well as a narrowed pore size), as opposed to those of pure MCM-48 and CMK-1 (**Table 2**), with the corresponding incorporated metallic species.

http://dx.doi.org/10.5772/63710

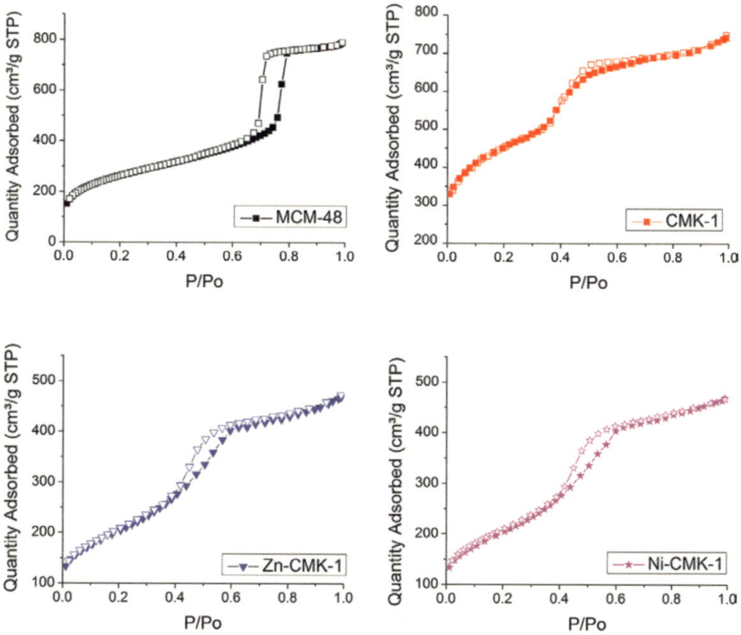

Figure 9. N_2 adsorption–desorption isotherms of MCM-48, mesoporous carbon CMK-1, Zn-CMK-1, and Ni-CMK-1.

Sample	Area (m^2/g)	VP ($cm^3\ g^{-1}$) mesoporous volume	BJH pore diameter (nm)	Metal cluster average size (nm)*	H_2 sorption (wt%)**
Si-MCM-48	1034	0.96	2.22	–	–
CMK-1	1089	0.69	3.11	–	2.1
Zn-CMK-1	568	0.55	2.65	Zn: 2.3 ZnO: 2.6	4.4
Ni-CMK-1	560	0.51	2.50	Ni Ni + NiO: 12	2.4

* Estimated by XRD (Scherrer formula) and TEM.
** H_2 sorption at 10 bar and 77 K.

Table 3. Textural and structural properties of the samples.

The sharp peak at pore size distribution of CMK-1 and Zn- and Ni-containing CMK-1 (**Figure 10**) indicates a fairly regular range of nanopores in carbonaceous materials, in correspondence with TEM observations.

The isotherm for CMK-3 and Pt-CMK-3 samples shows hysteresis loops at a relative pressure range of 0.4–0.8, which may be caused by capillary condensation–evaporation from the mesopores (**Figure 11**). The material appears in the form of fairly uniform carbon rods (see TEM images) with a higher pore volume as a result of the presence of mesopores between carbon rods. Irregularities (i.e., the presence of a fraction of nonlinear channels) can be attributed to SBA-15 meso-tunnels being incompletely filling with carbon precursor. The pore size distribution of CMK-3 shows a pronounced peak at 4 nm that points to a quite regular array of nanopores.

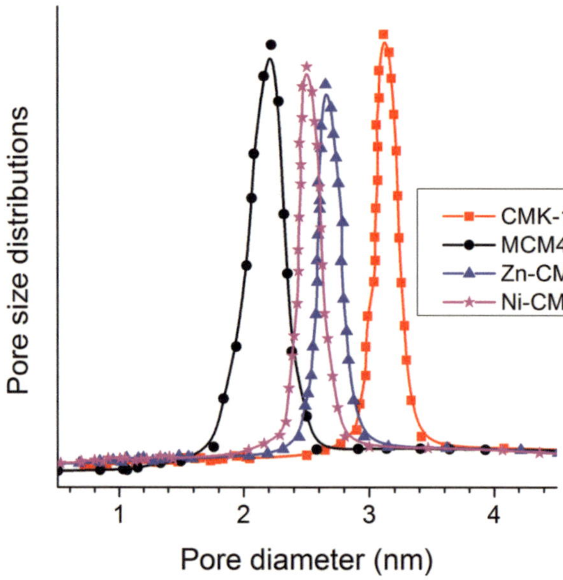

Figure 10. Pore size distribution of MCM-48, CMK-1, Zn-CMK-1, and Ni-CMK-1.

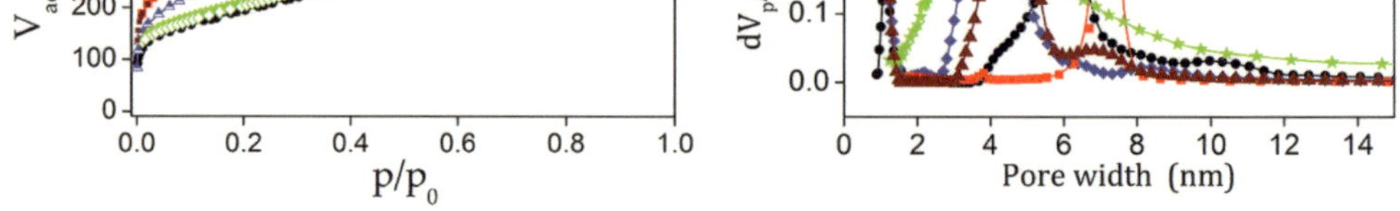

Figure 11. Nitrogen adsorption (solid symbols)–desorption isotherm (open symbols) at 77 K and pore size distribution of (a): SBA-15, (b) CMK-3, (c) Pt-CMK-3, (d) TiO_2-CMK-3, and (e) V-CMK-3.

On the contrary, the Pt-CMK-3 pore size distribution shows a broad peak with a maximum at approx. 3.8 nm, which is consistent with primary mesopores. The appearance of a wide peak can be ascribed to the filling of pores with Pt nanoclusters and the increase in the irregularity of nanopores. Around 6 nm an overlapping peak appears, corresponding to CMK-3 pores that are not filled with Pt nanoclusters. These analyses of pore size (see **Figure 11**) were consistent with observations from TEM.

Figure 11 also shows, the N_2 adsorption/desorption isotherms for SBA-15, CMK-3 and TiO_2-CMK-3 samples at 77 K; **Table 4** displays the textural properties determined from nitrogen physisorption analysis. The isotherm for SBA-15 is a typical type IV curve with hysteresis loop, according to IUPAC classification, revealing the mesoporous nature of the material.

The nitrogen adsorption/desorption isotherms for CMK-3 and TiO_2-CMK-3 are typical type IV curves exhibiting hysteresis loops type H_2, according to IUPAC classification, typical of mesoporous solids.

The materials obtained exhibit capillary condensation to relative pressures about 0.40 and 0.45, which can be related to the pore blocking effect of influencing pressure where evaporation/desorption pore occurs. The increase in adsorption at low relative pressures is associated with the presence of micropores or a strong adsorbate-adsorbent interaction.

TiO_2-CMK-3 reveals a reduced specific surface area, in comparison with those of pure SBA-15 and CMK-3, with incorporated anatase nanoparticles.

The CMK-3 pore size distribution shows a sharp peak at 4.2 nm that points to a quite regular array of nanopores. On the other hand, the pore size distribution of TiO_2-CMK-3 exhibits a broad peak with a maximum at approximately 6 nm, which is consistent with primary mesopores.

Material	SBET ($m^2 g^{-1}$)	VμP ($cm^3 g^{-1}$)	VTP ($cm^3 g^{-1}$)	Wp (nm)
Pt-CMK-3	500	0.03	0.98	3.8
V-CMK-3	1054	0.16	0.95	4.5
TiO_2–CMK-3	706	0.11	0.67	6.0
CMK-3	1323	0.23	1.01	4.3
SBA-15	1040	0.57	1.38	7.0

VTP, total pore volume; SBET, BET surface area; VμP, micropore volume; Wp, wide mesopore.

Table 4. Textural properties of the materials and composite.

N_2 adsorption/desorption isotherms and pore size distribution for CMK-3 and V-CMK-3 samples at 77 K shown **Figure 11** and **Table 4**, displays the textural properties determined from nitrogen physisorption analysis. The nitrogen adsorption–desorption isotherms for CMK-3 and V-CMK-3 are typical type IV curves exhibiting hysteresis loops type H_2, according to IUPAC classification, typical of mesoporous solids.

Materials obtained exhibit capillary condensation to relative pressures about 0.40 and 0.45, which can be related to the pore blocking effect of affecting the pressure where evaporation/desorption pore occurs. The increase in adsorption at low relative pressures of the samples

CMK-3 instead V-CMK-3 is associated with the presence of micropores or a strong adsorbate-adsorbent interaction.

V-CMK-3 reveals a reduced specific surface area, in comparison with those of pure CMK-3 (**Table 4**), with the corresponding incorporated metallic species.

5. XPS studies

5.1. Ni-CMK-1, Zn-CMK-1, Pt-CMK-3, and V-CMK-3 XPS characterization

Zn-CMK-1 XPS data exhibited that zinc appears as Zn^0 or Zn^{2+} ions (**Figure 12**), according to the binding energies (BE) of Zn ($2p_{3/2}$) electron at 1022.5 eV. The Zn $2p_{3/2}$ spectrum of Zn oxide, though not presenting multiple splitting or other complex effects, overlaps with Zn^0 species peak BE [29]. According to Woll [30], the Zn $2p_{3/2}$ lines for Zn and ZnO are quoted at 1021.4 eV and 1021.7 eV.

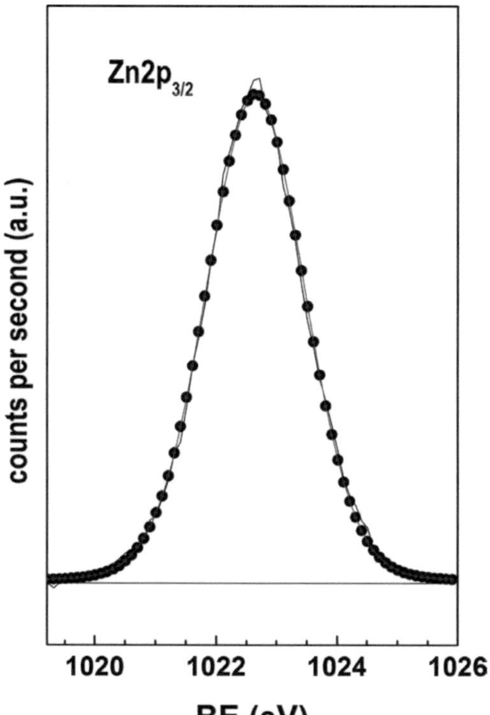

Figure 12. Zn-CMK-1 XPS spectrum.

The split 2p electron energy levels for Ni, Ni ($2p_{1/2}$), and Ni ($2p_{3/2}$) and seemed at 865–885 and 850–865 eV, respectively (**Figure 13**). These bond energies are lower by 1.5–2.0 eV for pure NiO [31]. This abnormality is possibly because the variance in organization of Ni^{2+} in NiO and on

http://dx.doi.org/10.5772/63710

CMK-1. It can confidently be said that Ni^{2+} ions are coordinated in higher symmetry in CMK-1, having a well-ordered crystal structure compared to that of NiO. The satellite band which appears at higher bond energy values also indicates the typical oxide structure.

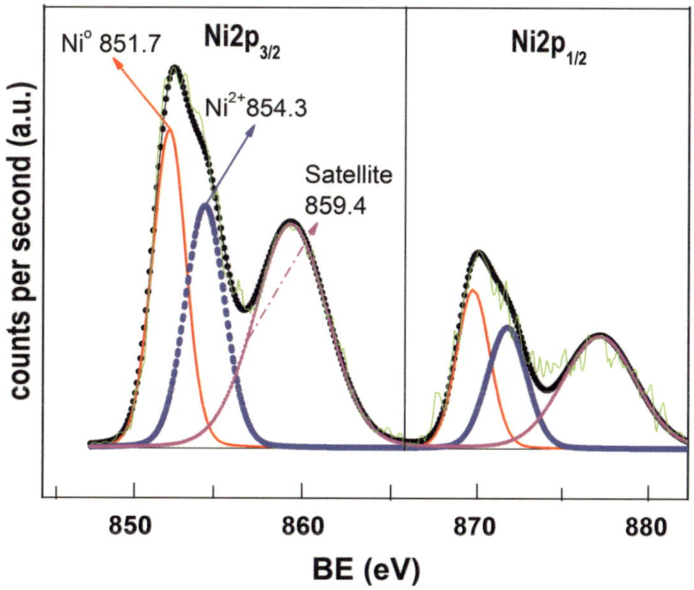

Figure 13. XPS binding energies spectra of the Ni $2p_{3/2}$ and Ni $2p_{1/2}$ peaks region separate by a vertical line. Gaussian component: Ni^0 solid line; NiO dot line and satellite peak of NiO dash dot line.

Activation under H_2 causes a partial reduction of the surface Ni^{2+} species, which yields Ni^0 (reduced catalyst). In addition to Ni^0, the reactive surface contains mixed Ni^0 + NiO. The mixed nickel compounds (such as nickel oxides and hydroxides) can appear in relatively large particles, where the NiO compound can be present on the outer surface, and metallic Ni is located in a the sub-surface close to the support, respectively. This distribution can be due to the drastic changes of the deposits' electronic structure that take place as the cluster size increases. Cluster size has a direct impact on the final state in the photoemission process because, once the process is over, the positive charge that remains on the aggregate can be less screened delocalized, compared with an extended metallic system. This entails a shift of the XPS spectrum corresponding to the Coulomb energy of localized charge. The shift proved to be proportional to the reciprocal particle diameter [32]. As a consequence, the binding energies of small metal aggregates should be found at higher values, as was in fact observed in a large number of cases. Elemental XPS analysis (Zn $2p_{3/2}$. O1s and C1s) showed that the chemical composition for Zn, O, and C, 0.2 %, 0.16%, and 99.64% (at.%), respectively, at 50–100 Å of depth for Zn-CMK-1 sample. Nevertheless, EDS study (**Figure 18c** and **d**) showed that the Zn content was 4.85 wt%, which suggests the Zn^0 (74%) lies inside the mesoporous of CMK-1 sample. Furthermore, by XPS, the oxygen concentration was around 80%, giving support to the knowledge that the ZnO lie on the external Zn-CMK-1 surface.

XPS study of Ni-CMK-1 point to a composition for Ni, O, and C as follows: 0.625, 0.375, and 99 (at.%), whereas by EDS analysis (5.3 and 0.6 wt% and for Ni and O, respectively). Therefore, a large amount of NiO is located on the outside of CMK-1 (>40%), consistent to the inferior quantities of Ni^0 create in Ni-CMK-1 mesoporous material. The EDS and XPS studies for Zn and Ni support the difference in cluster size obtained by XRD (Scherrer's formula) and TEM, shown in **Figure 18a** and **b** (2.5 nm and 12 nm, respectively). This is an indicator that the nanostructure of CMK-1 prevents the growth of large crystals in its nanopores, especially during the addition of zinc.

The XPS technique is normally used to determine the nature and oxidation state of Pt species (Pt^0, Pt^{2+}, and Pt^{4+}), and in particular the Pt (4f) peak study is employed (**Figure 14**). Metallic Pt^0 is known to have binding energies of 70.7–70.9 and 74.0–74.1 eV for $4f_{7/2}$ and $4f_{5/2}$ electrons, respectively [33]. In oxidized states, the binding energies exhibited by Pt^{2+} and Pt^{4+} are much higher: 72.8–73.1 eV ($4f_{7/2}$) and 76.3–76.4 eV ($4f_{5/2}$) for Pt^{2+} and 74.6–74.9 eV ($4f_{7/2}$) and 78.1–78.2 eV ($4f_{5/2}$) for Pt^{4+} [34].

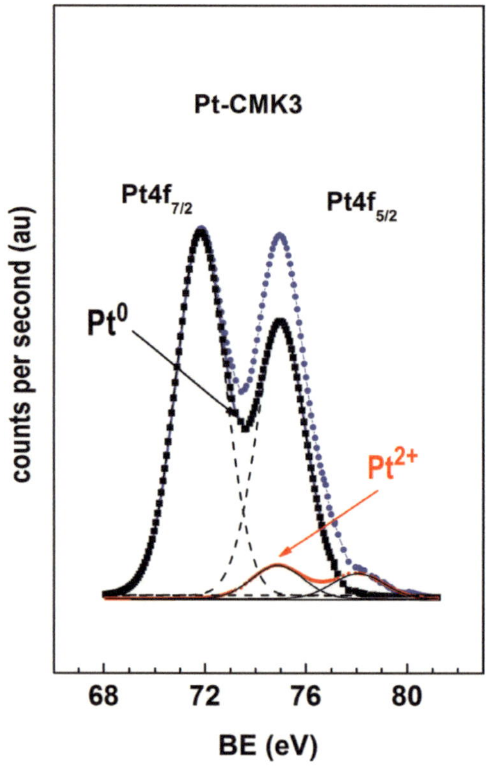

Figure 14. Core level Pt4f XPS spectra of Pt-CMK-3.

The spectra showed two pairs of peaks. As we can see all spectra displays signals at low and high energy (doublets), Pt $4f_{7/2}$: 71.8 eV and $4f_{5/2}$: 75.1 eV, demonstrating that Pt nanoclusters are metallic (**Figure 14**). The Pt^0 peak moves slightly to higher binding energies; this phenomena is in agreement to the characteristic effect of small particle sizes, as published by Roth et al. [34] and Takasu et al. [35] or for charge transfer between carbon and platinum described by Aricò et al. [36]. The bands that appeared in the second pair of platinum signs, on the other hand, appears around 74.9 and 77.1 eV, which has been reported in the literature to be due to the presence of PtO or PtO_2 [37].

According to the XPS analysis, the Pt concentration at 50 Å of depth was 0.38 wt%, whereas ICP and EDS analyses pointed to a nominal Pt = 0.9 wt% on Pt-CMK-3. This led to the suggestion that the majority of Pt^0 lies inside the nanostructure of the CMK-3 sample. Moreover, about 80 wt% of oxygen was on the outer surface of CMK, with the correspondingly lower amount of not reduced Pt species ($Pt^{2,4+}$).

Figure 15 shows the binding energy for V $2p_{3/2}$core level of XPS spectra for V-CMK-3. The only one contribution located about 515.0–515.2 eV is assigned to V^{3+} in the form of V_2O_3. Vanadium oxidation states is consistent with XPS spectra previously taken of V_2O_3, which we attributed to $2p_{3/2}$ core electrons [38]. The peak at 525 eV is consistent with the $2p_{1/2}$ peak of vanadium [38].

Vanadium 3^+ has a d^2 electronic configuration and, from inspection of the Tanabe-Sugano diagram for a d^2 ion [39], is expected to have three spin-allowed ground-state absorption transitions in both tetrahedral and octahedral coordination. Therefore, 3^+ is a possible oxidation state of vanadium in CMK-3 according to XPS data.

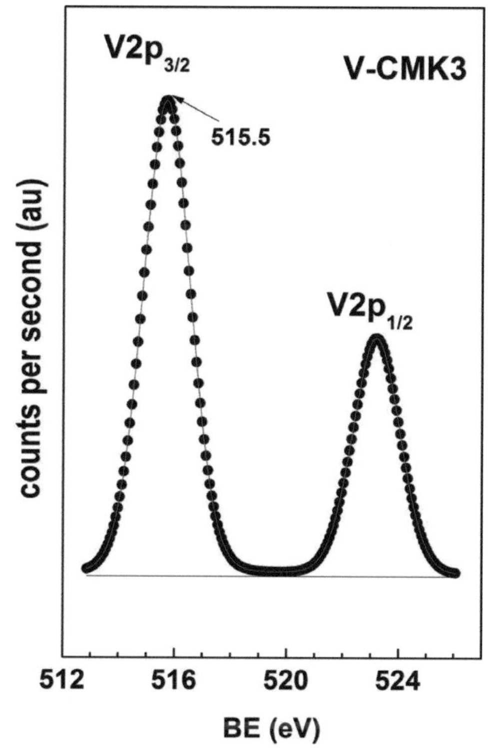

Figure 15. Core level V2p XPS spectra of V-CMK-3.

6. TiO_2-CMK-3 Raman studies

Figure 16 shows Raman spectra of SBA-15, pristine host CMK-3, TiO_2-CMK-3, and pure anatase and rutile in the range of 200–1200 cm^{-1}.

For SBA-15, Raman bands at 430, 800 and 1080 cm^{-1} are observed. The bands at 430 and 1080 cm^{-1} can be attributed to symmetric and asymmetric vibrations of the Si-O-Si unit, respectively. The band at 800 cm^{-1} is the symmetric stretching mode of the tetrahedral [SiO_4] unit [40–42]. **Figure 16** also shows the spectra of TiO_2 onto CMK-3; the characteristic signals of anatase are detected in the sample if compared with those of pure anatase [41,43,44] and pure rutile.

In inset **Figure 16**, we illustrated the Raman spectra of bare CMK-3 and TiO_2-CMK-3 (from 1100 to 1700 cm^{-1}). The Raman spectrum of the obtained CMK-3 as well as of TiO_2-CMK-3 shows two peaks, at 1578 and 1353 cm^{-1}. The peak at 1578 cm^{-1} corresponds to an E_2g mode of graphite and is called G band, related to C═C-double-bonded carbon vibration in hexagonal lattice, while the peak at 1353 cm^{-1} is associated with vibrations of C-C bond, referred to as D band, attributed to the configuration of disordered graphite [45–47]. In the Raman spectrum of graphitic materials such as activated charcoal, carbon black, and CMK, this second line

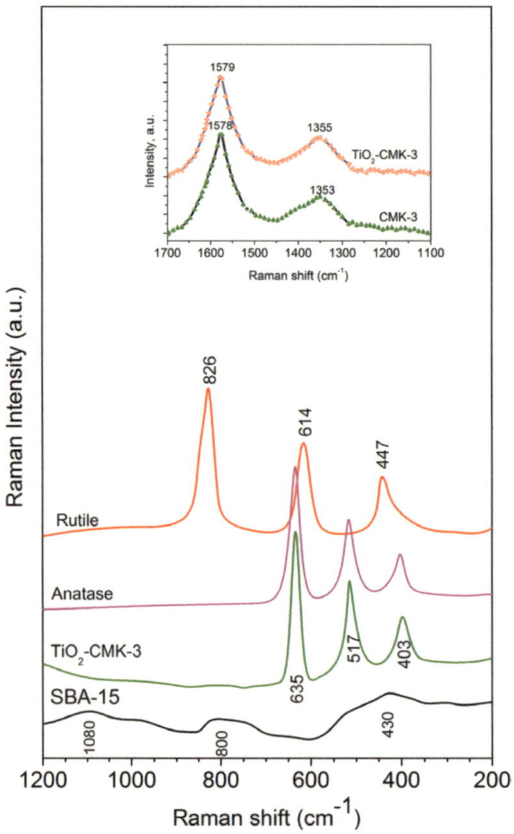

Figure 16. Raman spectra of SBA-15, TiO_2–CMK-3, rutile, anatase, and CMK-3 (inset).

appears. The relative intensity of the two lines depends on the type of graphitic material. The intensity of 1353 cm^{-1} line increases with an increase in carbon in the samples and with a decrease in the graphite crystal size. Hence, a signal at 1355 cm^{-1} might possibly be ascribed to a diamond-like atomic arrangement in the graphite samples, probably as partially tetrahedral bonded [48]. This agrees with the proposal of the graphite-like structure for CMK-3 [49]. It could be clearly seen that G and D band intensity of TiO_2/CMK-3 was not perturbed significantly (inset **Figure 16**).

7. TEM studies

Figure 17 shows TEM images of CMK-1. Ordered structure was shown to be slightly damaged by thermal treatments.

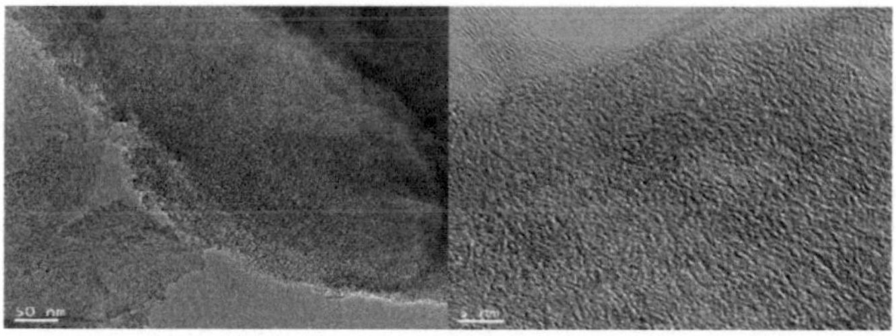

Figure 17. TEM images of CMK-1.

The TEM micrographs of Ni-CMK-1 and Zn-CMK-1 were depicted in **Figure 18**, showing. The ordered cubic Ia3d mesostructure of Metal-CMK-1 (**Figure 18a** and **b**), indicating that the ordered structure of CMK-1 is retained after the incorporation of Ni and Zn nanoparticles.

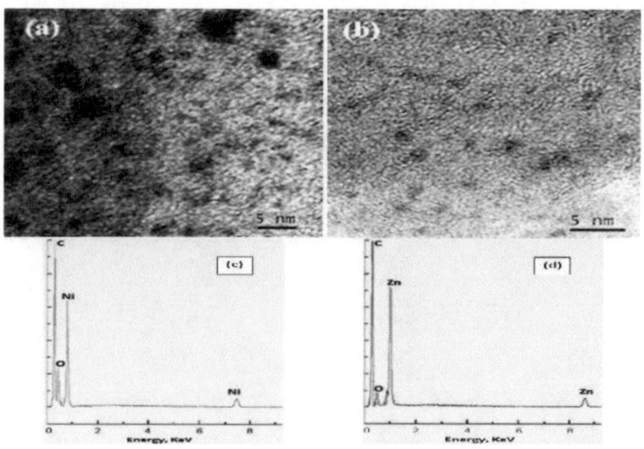

Figure 18. TEM images of (a) Ni-CMK-1 and (b) Zn-CMK-1. (c and d) EDS spectrum of: (c) Ni-CMK-1 and (d) Zn-CMK-1.

Figure 19. TEM images of SBA-15. (a) [1 0 0] and (b) [1 1 0] reflection planes.

The larger dark areas over the channels most likely correspond to NiO nanoparticle agglomerates on the external surface with average diameter of 11–13 nm (**Figure 18a**). The small dark spots in the image (**Figure 18b**) could be ascribed to Zn and ZnO nanoparticles with average diameter of ~2.5–3 nm probably located into the host pores. EDS analysis was performed to estimate the presence of Zn- and Ni-containing CMK-1. A careful evaluation of the elemental compositions for C, O, and Zn and Ni nanoparticles was performed (**Figure 18c** and **d**).

According to the maps collecting five spots, the average atomic weight % of C, O and Zn nanoclusters in Zn-CMK-1 is about 94.94, 0.26, and 4.80%, respectively. This shows that only 20% of the Zn species appear as ZnO. On the other hand, the average atomic weight % for Ni-CMK-1 was 94.1, 0.6, and 5.3 wt% for C, O, and Ni, clearly indicating that the higher Ni content appears as oxide (about 40%).

Figure 19a shows TEM images of synthesized SBA-15 used as an inorganic template. In **Figure 19b**, an ordered mesoporous array of longitudinal nanochannels can be seen, with pore

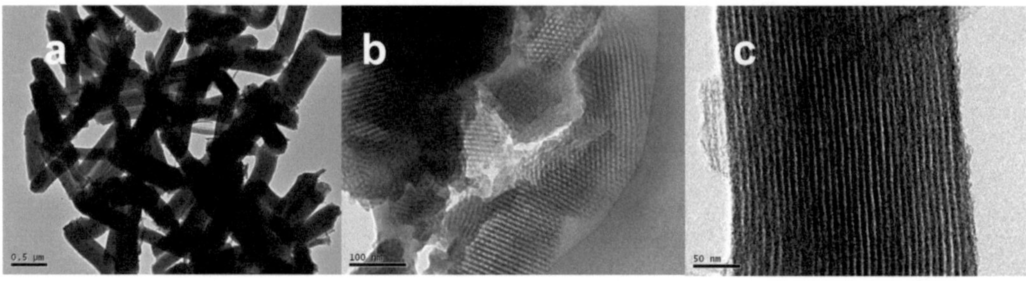

Figure 20. TEM images of CMK-3. (a) Nanoparticles (b) [1 0 0] and (c) [1 1 0] reflection planes.

diameter of 7.5 nm. TEM images of CMK-3 are shown in **Figure 20**. Ordered structure proved to be slightly damaged by thermal treatments and exhibits well-organized pores parallel to one another. The white lines correspond to the mesopores generated in the space where the walls of the SBA-15 template had been.

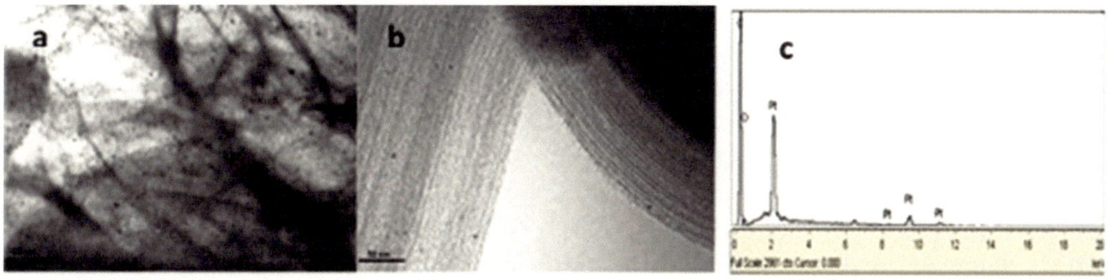

Figure 21. TEM images (a, b) and EDS (c) spectrum of Pt-CMK-3.

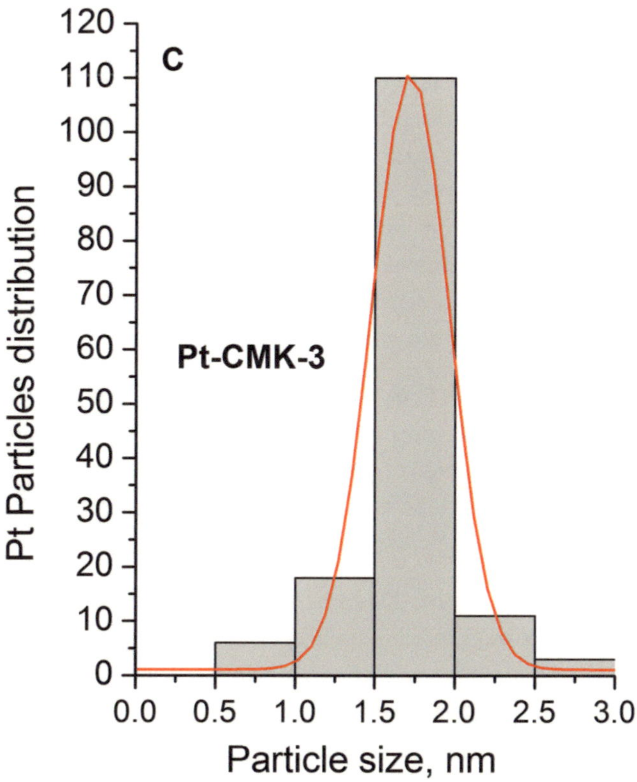

Figure 22. Histograms of PSDs measured from full images of Pt-CMK-3 shown in **Figure 21**.

Figure 21 offers an illustration of TEM micrographs of Pt-CMK-3. The ordered mesostructure of the sample can be seen in **Figure 21a** and **b**, indicating that the ordered structure of CMK-3 persists after Pt nanoparticles are incorporated. The darker contrast (small dark spots) which can be seen in some places may be attributed to the presence of regularly dispersed Pt particles. These images evidence the fine dispersion of Pt^0 on CMK-3, a tinny size dissemination (**Figure 22**), and mean diameter of ~1 to 2 nm inside the mesopores of CMK-3 nanostructure. The existence of Pt^0 inside the pore system of CMK-3 indicates an enhancement in the dispersion phenomena compared with the metal on the external carbon surface. The Pt nanoclusters are active sites for hydrogen adsorption and narrow pore size distribution is vital because the level of dispersion is better and the metal active area is superior.

The metal particles dispersed in CMK-3 carbon have from 1 to 2.5 nm size, and the average particle diameter are approximately to 1.74 nm, showing that utmost of Pt nanoclusters are mainly in the nanostructured intra-channel system of CMK-3.

Dispersion and size distribution of metal particles of Pt toughly depends on the technique of deposition and the characteristics of carbon support as published by Kuppan and Selvam [25], suggesting that reduction with paraformaldehyde of platinum nanoparticles deposited on mesoporous carbon CMK-3 is superior to other methods, including H_2 reduction. In this work, we show that the nature of Pt incorporation to CMK-3 and the preparation of Pt-CMK-3 before Pt reduction (under N_2 atmosphere and controlled temperature removal of the Pt precursor) impact the subsequent size of Pt-reduced nanoparticles, resulting in a narrow distribution of around 1.74 nm (**Figure 22**). If the first elimination of Pt precursors-CMK-3 occurs in an oxidative atmosphere, the likelihood of generating larger Pt clusters increases. Similar consequence on dispersion of iridium over SBA-16 was reported recently [37]. The factor of the Pt source removal (chloroplatinic acid in ethanol-CMK-3) is endothermic and not oxidative under N_2 flow and controlled temperature desorption, avoiding the relocation and accumulation of platinum particles after it reduction.

In order to determine the presence of Pt in the prepared Pt-CMK-3, EDS analysis was employed to carefully evaluate the elemental compositions for C, O, and Pt nanoparticles (**Figure 21c**). Maps of elements were collected from four spots and used to evaluate the presence of all elements. This showed that the average atomic percentage of C, O, and Pt nanoclusters in Pt-CMK-3 is about 98.82, 0.28, and 0.9%, respectively.

TEM images for TiO_2-CMK-3 (**Figure 23**) indicated an ordered structure slightly damaged by thermal treatments; they also exhibit well-organized pores parallel to each other. The white lines correspond to the mesopores generated in the space previously occupied by the walls of SBA-15 template. In **Figure 23b**, a dark spot indicates anatase nanoclusters. **Figure 23c** shows a TEM image with negative contrast, displaying nanoclusters with bright spots, allowing the observation of nanoclusters. Particle size distribution ranged from 1 to 11 nm and mean particle diameter was found to be close to 5 nm, indicating that most of the particles reside within of the pore system.

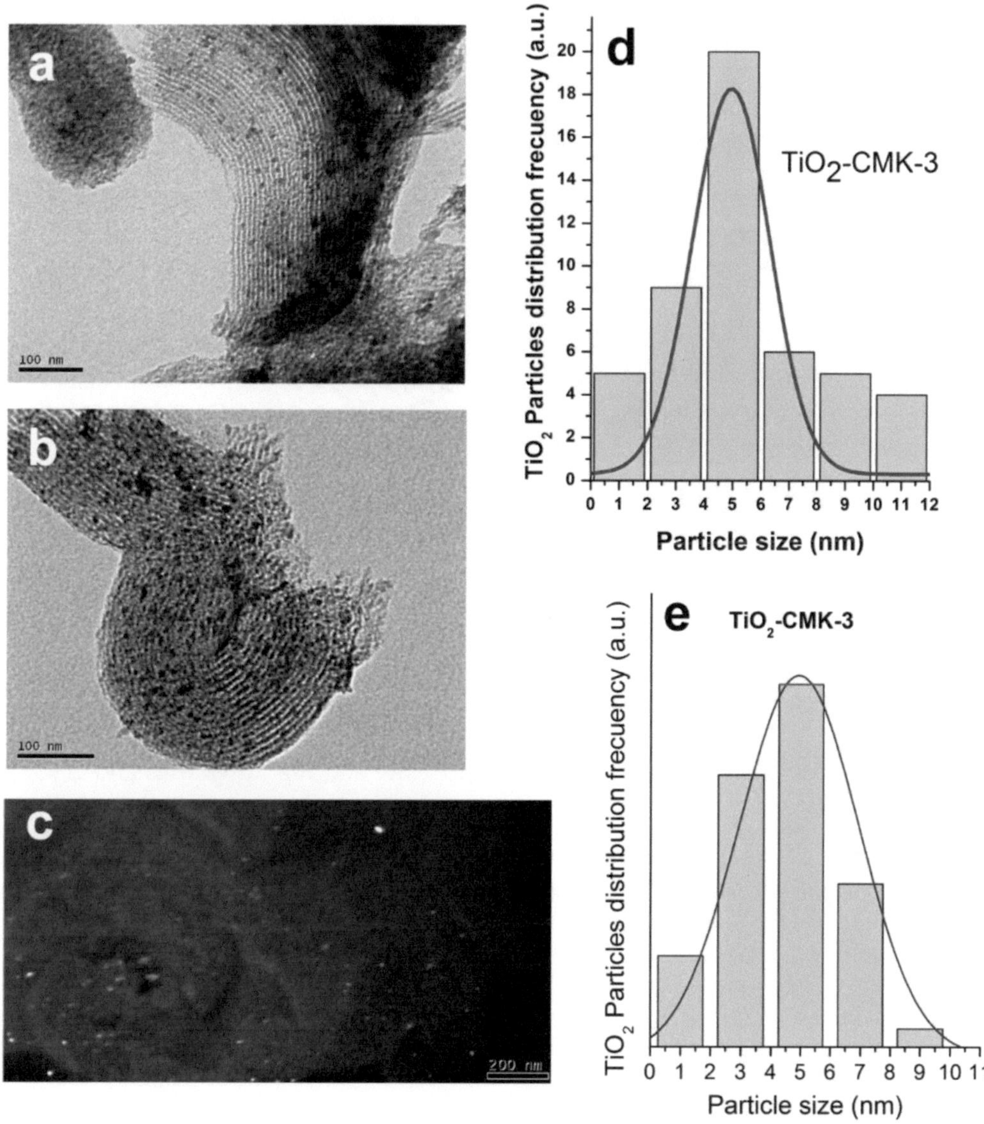

Figure 23. (a) and (b) TEM images TiO_2–CMK-3 and (c) Negative contrast image. (d) and (e) particle size distribution histogram corresponding to (a) and (c) TEM images, respectively.

In this work, we show that the nature of TiO_2 incorporation into CMK-3 and the preparation of TiO_2-CMK-3 (under N_2 atmosphere and controlled temperature elimination of Ti precursor), influence posterior TiO_2 nanoparticle size, leading to a narrow distribution as shown in the histograms (**Figure 23a** and **c**).

Vanadium nanoclusters were not seen in TEM images of V-CMK-3 due to a low contrast in the TEM image (**Figure 24**).

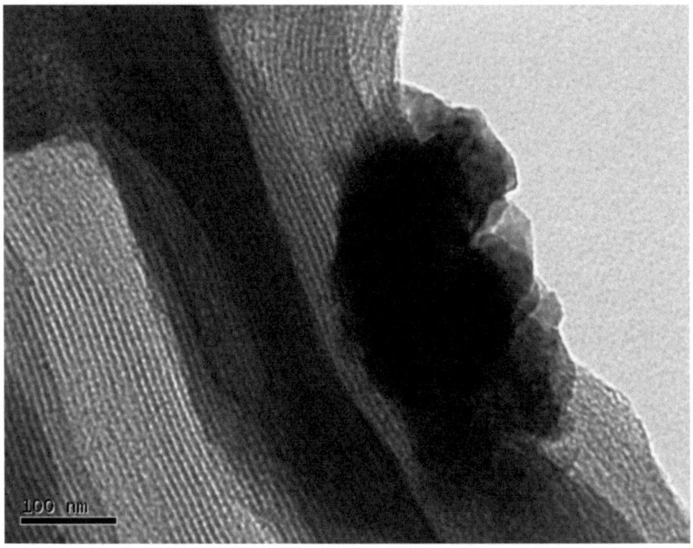

Figure 24. TEM images of V-CMK-3.

8. Hydrogen uptake measurements

The capacity of hydrogen storage was evaluated at low and high pressures and cryogenic temperatures (77 K).

The experimental data were fitted by Freundlich isotherm [50], that is, a purely empirical formula for gaseous adsorbates. The Freundlich model (Eq. (2)) is described by the following equation:

$$Q = K_F\, P^{\frac{1}{n}} \tag{2}$$

Here, Q is the quantity per unit mass of adsorbent, P is the adsorbate pressure, and KF and n are experiential constants (Freundlich constants) for, respectively, adsorbent-adsorbate pair at a constant temperature. A measure of the adsorption intensity is characterized by n, and K_F is the adsorption measurement [50]. This equation was applied to the experimental data using nonlinear least-square regression for arbitrary fit functions and minimizing the objective function by means of the Levenberg-Marquardt method. The fitting accuracy was $R^2 = 0.98$.

The hydrogen storage is defined as "uptake," the carbon material behaves as a sponge, due to the fact that the process is completely reversible and the curve remains in the same place as each point returned to baseline when measurements were carried out without the bath of liquid N_2. We performed the treatment at high temperatures (except TiO_2-CMK-3 sample where treatment is done only under N_2 flow) is to reduce the oxide cluster to a metallic cluster because

http://dx.doi.org/10.5772/63710

the electronic configuration of the metal cluster interacts better with hydrogen molecules and for high H_2 storage. The amount of hydrogen uptake of all samples is higher than carbon samples. The Zn-CMK-1 sample has the highest hydrogen uptake (0–10 bar). Yet, at high pressures, the sample with Zn shows greater H_2 adsorption than all the other samples. In samples that contain metal clusters, there are at least two ways in the process of hydrogen physisorption on metal-CMK-1/CMK-3 surface. Hydrogen molecules spill over onto the CMK-1 nanopores/micropores and are adsorbed onto nanometric metal clusters. Hydrogen adsorption proved to be fully reversible, which indicates that there was no chemical reaction or strong bond between hydrogen and metal nanocluster or the CMK-1 framework.

The proposed mechanism for hydrogen storage in metal-CMK-1/CMK-3 agreed with Kim and Park [51]. Hence, a dipole-induced model is proposed. The first layer of hydrogen molecules could react with the metal cluster because of the high oxidation capacity of metals interacting as dihydrogen complex [35], but the level of this initial interaction is not significant. The second layer of hydrogen molecules is physically adsorbed by dipole-induced interaction. Hydrogen molecules are basically non-polar, but the strong interaction of the metal particles leads to the dipole-induced effects of the hydrogen molecules. The third layer and any other layer above the hydrogen molecules could interact with metal cluster by the same mechanism, but the force of the dipole-induce bond is weak and decreases in inverse proportion with the distance from the surface. Consequently, this mechanism can be applied at higher pressures, since metal-CMK-1/CMK-3 adsorbs a larger amount of hydrogen than carbon at higher pressures.

Theoretically, the 3d states of metals mainly contributed to ground-state electronic structure near the Fermi level of metal-CMK-1 and, to some degree, the p states of carbon. Those contributions of 3d states decrease as we move across the periodic table from Mn to Zn following the increase of d orbital occupancies [52].

Hydrogen bonds to the metal result in an opening of HOMO-LUMO energy gap in the case of Ni-CMK-1, while it shows a reduction in HOMO-LUMO energy gap for Zn-CMK-1, resulting in the appearance of molecular electronic states which allow a better H_2 physisorption.

The understanding at an atomic scale of the hydrogen spillover mechanism for storage of hydrogen in metal-doped carbon materials and metal-organic frameworks is discussed by means of a critical assessment of recent computational and experimental studies. It is claimed that the spillover mechanism includes (a) generation and desorption of mobile H atoms on metal nanoparticles; (b) diffusion of H atoms in weakly bound states on the support; (c) sticking and immobilization of H atoms at preferential locations of the receptor where barriers to sticking are lower, and (d) Eley-Rideal recombination of the adsorbed H atoms with diffusing mobile H atoms to form H_2 [53].

In the TiO_2-CMK-3, surface can adsorb hydrogen molecules by spill over onto CMK-3 nanopores/micropores and adsorbed onto nanometric anatase clusters. First, we could attribute this adsorption to the different particle surface areas, since the adsorbed quantity might be primarily related to the available free surface area. Second, we could attribute it to the fact that nanometric carbon promoting the high-surface areas of TiO_2 nanoclusters. CMK-3 surface promotes the high-surface areas of TiO_2 nanoclusters may possibly modify oxidation state and control porosity feature, which is important to cause hydrogen adsorption (H_2 molecules) by weak chemical pathways (such as dihydrogen complex interaction), although anatase or rutile is generally weak in hydrogen storage [54]. Through XRD and TEM observation, we also noticed that with the assistance of CMK-3, the fabricated TiO_2 nanoparticles show small size and high dispersion, contributing to the substantial increase in H_2 adsorption. In other words, the presence of TiO_2 nanoclusters in CMK-3 results in formation of better-localized states of electrons at TiO_2 nanoclusters and CMK-3 interface. These localized states may distribute electron concentration at the surface of TiO_2-CMK-3 composite and causes a few reactions between hydrogen molecules and surface TiO_2 nanoclusters [54]. According to theoretical investigations [55], TiO_2 and CMK-3 interaction is not affected by the physisorption of the first H_2 molecule because it does not impact the distance between Ti and the nearest C. The H-H distance is influenced (elongated) by bonds Ti-C sp2d because the strong repulsive phenomena on the adsorbed H_2. The hydrogen atoms of the adsorbed H_2 molecule should have negative charges, and the H_2 molecule can be regarded as trapped by Ti cation by means of the charge polarization mechanism. This is an indicator that Ti donates electrons to the nearest C atoms on CMK-3, where the d-orbitals of the Ti atom overlap the sp^2 orbitals of the Ti-C bonds to form the mixed sp^2d hybridization. This charge transfer is possible because the Ti atom is in cationic form and renders extensive hetero-polar bonding between the Ti atom and its neighboring C atoms, resulting in an increase in H_2 molecule uptake. Moreover, CMK-3 approaching the positively charged Ti cation leads to loss of d-orbital degeneracy since the electrons of CMK-3 will be closer to some d-orbitals while being farther away from others. Thus, those d-orbitals that are closer to CMK-3 have a higher degree of energy than those farther away. This results in d-orbitals splitting in energy to reduce the total energy and stabilize the system, as explained by the ligand field theory [56].

Therefore, the first layer of hydrogen molecules could be reacted with the metal cluster due to high oxidation/reduction metals capacity interacting like as dihydrogen complex [35], similar metal-carbon samples; yet, the amount of this initial interaction is negligible. The second layer of hydrogen molecules is physically adsorbed by dipole-induced interaction as metal-carbon samples. Hydrogen molecules are non-polar, but the strong interaction of anatase particles leads to dipole-induced effects in H_2. The third level of the layers and any higher layer of H_2 molecules could bond with metal cluster by the same way; however, the dipole-induced interaction is weak, decreasing as separation to the surface increases. At higher pressure, this mechanism can be applied; consequently, the adsorption of hydrogen to TiO_2-CMK-3 is higher than CMK-3 at greater pressures. The understanding of the distribution of frontier molecular orbitals around a nanostructure would be a significant guideline to design new functionalized materials for hydrogen storage. While strong localization of the highest occupied molecular orbital (HOMO) occurs on Ti cation located at the surface on anatase nanocluster, strong delocalization of the lowest unoccupied molecular orbital (LUMO) occurs at the base area. The strong localization of HOMO on Ti cation at the surface explains the enhanced adsorption interactions of H_2 molecules [55]. Moreover, hydrogen bonds to anatase result in an opening of HOMO-LUMO energy gap. A reduction in HOMO-LUMO energy gap for TiO_2-CMK-3 results in the appearance of molecular electronic states that allow a better H_2 physisorption [52].

The hydrogen storage on V-CMK-3 sample could be attributed to a high V_2O_3 particle surface areas and could be attributed to the fact that nanometric carbon helping the high-surface areas of V_2O_3 nanoclusters. However, the high hydrogen adsorption may be explain because hydrogen storage might increase via the spill-over phenomenon. Specifically, hydrogen molecules are dissociated on the surface of the introduced transition metal cation (V^{+3}). The other potential explanation is that the hydrogen molecules are attracted into the graphite pores due to formed weak dipoles with V_2O_3 nanoclusters. Metal cation increases and quickens hydrogen storage through a weak chemical reaction (from either the spill-over phenomenon or dipole attraction) (**Figure 25**).

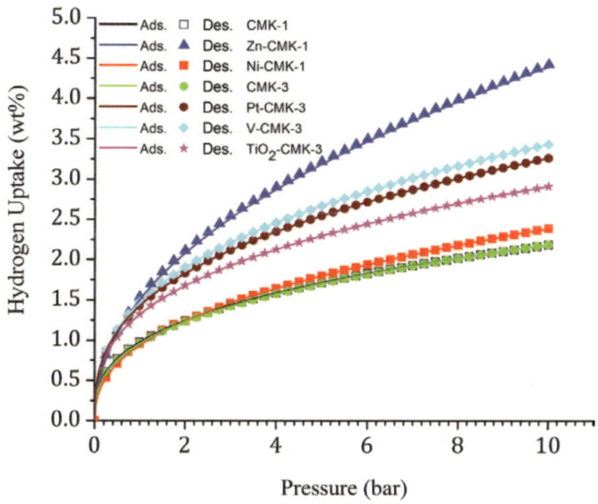

Figure 25. Isotherms of hydrogen at 77 K on the different materials studied at a range of pressures (0–10 bar).

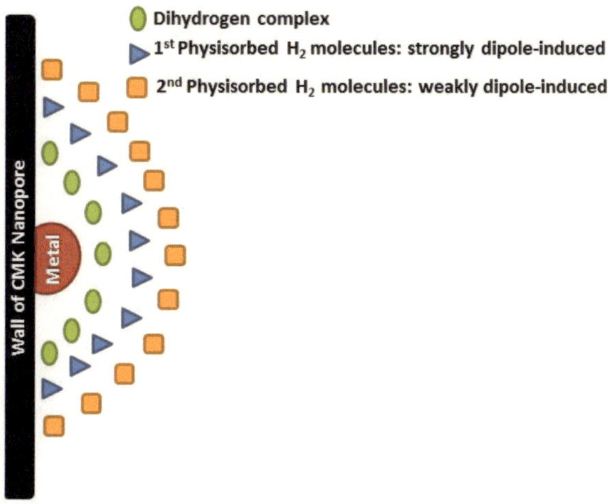

Figure 26. Scheme for the proposed mechanism of metal-CMK and hydrogen interaction.

The scheme showed in **Figure 26** suggests a mechanism for hydrogen storage in carbon mesoporous CMK, modified with metals in agreement with Kim et al.

9. Conclusions

We have shown that a hopeful hydrogen storage material can be obtained by ordered porous carbons CMK-1/CMK-3 modified with Ni, Zn, Pt, Ti, and V species that were synthesized by replication using MCM-48 and SBA-15 as hard template, respectively. The metal-metal/oxide nanoparticles-containing carbon was obtained by wetness impregnation considering the thermal treatments employing H_2. Zn species mostly found as metallic Zn while Ni as NiO, with the highest average cluster size (12 nm), indicating greater dispersion of zinc and its incorporation in the nanoporous of mesoporous carbons but without occlusion by their small size compared with NiO clusters. The higher Zn dispersion probably causing a high metal surface area and a better use of the support allowed to a superior performance to H_2 uptake of Zn-CMK-1. Pt-CMK-3 sample present an improvement in the capacity of hydrogen uptake than CMK-3 carbon due to the better dispersion of uniform platinum nanoparticles as well as better utilization of the support, which may probably was originate of a high-surface area and pore volume, that allow a large dispersion of Pt. Pt nanoparticles in Pt-CMK-3 have smaller size (~1.7 nm average) and have a quite well dispersion inside the nanochannels of CMK-3. This shows the better hydrogen uptake of the Pt-CMK-3 than CMK-3 carbon.

TiO_2-CMK-3 can significantly enhance hydrogen adsorption capacity and hydrogen storage performance of CMK-3 material, proving to prospective candidates for application in hydrogen storage. CMK-3 modified with TiO_2 in anatase phase shows an enhanced capacity for H_2 storage than that of the nanoporous CMK-3 sample. TiO_2-CMK-3 improved performance and larger activity for H_2 uptake is attributed to improved dispersion of uniform anatase nanoparticles in addition to efficient use of the CMK-3, probably originate in the high pore volume and surface area, allowing a great dispersion of anatase. TiO_2 (anatase phase) nanoparticles in TiO_2-CMK-3 have a smaller size (~5 nm average, obtained by XRD and TEM) and a quite good dispersion.

Vanadium-modified carbon sample has higher hydrogen storage than CMK-3 pristine material and is the best material for hydrogen adsorption using CMK-3 nanometric carbon as support but lower than Zn-CMK-1.

A hydrogen uptake mechanism on metal/carbon surfaces was proposed. We can conclude by indicating that CMK-1/CMK-3 hydrogen storage capacity was increased by addition of metal clusters. Moreover, hydrogen storage behaviors onto metal/metal oxides-carbons can be optimized by controlling the metal cluster size and dispersion and by increasing the carbon-specific surface area.

http://dx.doi.org/10.5772/63710

Acknowledgements

Juliana M. Juarez, Marcos Gomez Costa, Oscar A. Anunziata, NANOTEC, CONICET, UTN-FRC, Maestro Lopez y Cruz Roja Argentine, Cordoba Argentine. The authors thank to CONICET Argentina, PIP CONICET 11220120100218CO. 2014–2016.

Author details

Marcos B. Gómez Costa, Juliana M. Juárez and Oscar A. Anunziata*

*Address all correspondence to: oanunziata@scdt.frc.utn.edu.ar

Centre for Research in Nanoscience and Nanotechnology (NANOTEC), Cordoba Regional Faculty, National Technological University, Maestro López y Cruz Roja Argentine, Córdoba, Argentina

References

[1] Schlapbach L, Züttel A. Hydrogen-storage materials for mobile applications. Nature. 2001; 414: 353–358.

[2] Kajiura H, Tsutsui S, Kadono K, Kakuta M, Ata M, Murakami Y. Hydrogen storage capacity of commercially available carbon materials at room temperature. Applied Physics Letters. 2003; 82: 1105–1107.

[3] Mandoki NT, Dentzer J, Piquero T, Saadallah S, David P, Guterl CV. Hydrogen storage in activated carbon materials: role of the nanoporous texture. Carbon. 2004; 42: 2744–2747.

[4] Züttel A, Sudan P, Mauron P, Kiyobayashi T, Emmenegger C, Schlapbach L. Hydrogen storage in carbon nanostructures. International Journal of Hydrogen Energy. 2002; 27: 203–212.

[5] Fang G, Liu G, Yang Y, Wang S. Quartz crystal microbalance sensor based on molecularly imprinted polymer membrane and three-dimensional Au nanoparticles@mesoporous carbon CMK-3 functional composite for ultrasensitive and specific determination of citrinin. Sensors and Actuators B. 2016; 230: 272–280.

[6] Gadiou R, Saadallah SE, Piquero T, David P, Parmentier J, Guterl CV. The influence of textural properties on the adsorption of hydrogen on ordered nanostructured carbons. Microporous and Mesoporous Materials. 2005; 79: 121–128.

[7] Fang BZ, Zhou HS, Honma I. Ordered porous carbon with tailored pore size for electrochemical hydrogen storage application. The Journal of Physical Chemistry B. 2006; 110: 4875–4880.

[8] Zhai Y, Dou Y, Zhao D, Fulvio PF, Mayes RT, Dai S. Carbon materials for chemical capacitive energy storage. Advanced Materials. 2011; 23: 4828–4850.

[9] Yang ZX, Xia YD, Sun XZ, Mokaya R. Preparation and hydrogen storage properties of zeolite-templated carbon materials nanocast via chemical vapor deposition: effect of the zeolite template and nitrogen doping. The Journal of Physical Chemistry B. 2006; 110: 18424–18431.

[10] Wegrzyniak A, Jarczewski S, Wach A, Hedrzak E, Kustrowski P, Michorczyk P. Catalytic behaviour of chromium oxide supported on CMK-3 carbon replica in the dehydrogenation propane to propene. Applied Catalysis A: General. 2015; 508: 1–9.

[11] Anbia M, Parvin Z. Desulfurization of fuels by means of a nanoporous carbon adsorbent. Chemical Engineering Research and Design. 2011; 89: 641–647.

[12] Juárez JM, Gómez Costa MB, Anunziata OA. Synthesis and characterization of Pt-CMK-3 hybrid nanocomposite for hydrogen storage. International Journal of Energy Research. 2015; 39: 128–139. doi:10.1002/er.3229.

[13] Juárez JM, Gómez Costa MB, Anunziata OA. Preparation and characterization of activated CMK-1 with Zn and Ni species applied in hydrogen storage. International Journal of Energy Research. 2015; 39: 941–953. doi:10.1002/er.3298.

[14] Benhamou A, Basly JP, Baudu M, Derriche Z, Hamacha R. Amino-functionalized MCM-41 and MCM-48 for the removal of chromate and arsenate. Journal of Colloid and Interface Science. 2013; 404: 135–139.

[15] Jorda-Beneyto M, Lozano-Castello D, Suarez-Garcia F, Cazorla-Amoros D, Linares-Solano A. Advanced activated carbon monoliths and activated carbons for hydrogen storage. Microporous and Mesoporous Materials. 2008; 112: 235–242.

[16] Valles VA, Ledesma BC, Rivoira LP, Cussa J, Anunziata OA, Beltramone AR. Experimental design optimization of the tetralin hydrogenation over Ir–Pt-SBA-15. Catalysis Today. 0920–5861. doi:10.1016/j.cattod.2015.07.018. Article in press. Available on line at http://www.sciencedirect.com/science/article/pii/S0920586115004095

[17] Ryoo R, Joo SH, Jun S. Synthesis of highly ordered carbon molecular sieves via template-mediated structural transformation. Journal of Physical Chemistry B. 1999; 103:7743–7746.

[18] Solovyov LA, Zaikovskii VI, Shmakov AN, Belousov OV, Ryoo R. Framework characterization of mesostructured carbon CMK-1 by X-ray powder diffraction and electron microscopy. Journal of Physical Chemistry B. 2002; 106: 12198–12202.

[19] Huwe H, Fröba M. Iron (III) oxide nanoparticles within the pore system of mesoporous carbon CMK-1: intra-pore synthesis and characterization. Microporous and Mesoporous Materials. 2003; 60: 151–158.

[20] Suryavanshi U, Iijima T, Hayashia A, Hayashi Y, Tanemura M. Fabrication of ZnO nanoparticles confined in the channels of mesoporous carbon. Chemical Engeneering Journal. 2012; 179: 388–393.

[21] Habibi MH, Rahmati MH. Fabrication and characterization of ZnO@CdS core–shell nanostructure using acetate precursors: XRD, FESEM, DRS, FTIR studies and effects of cadmium ion concentration on band gap. Spectrochimica Acta Part A: Molecular and Biomolecular Spectroscopy. 2014; 133: 13–18.

[22] Chen R, Han J, Yan X, Zou C., Bian J, Alyamani A, Gao W. Photocatalytic activities of wet oxidation synthesized ZnO and ZnO–TiO_2 thick porous films. Applied Nanoscience. 2011; 1: 37–44.

[23] Gómez Costa MB, Juárez JM, Martínez ML, Beltramone AR, Cussa J, Anunziata OA. Synthesis and characterization of conducting polypyrrole/SBA-3 and polypyrrole/Na-AlSBA-3 composites. Material Reseach Bulletin. 2013; 48: 661–667.

[24] Veena Gopalan E, Malini KA, Santhoshkumar G, Narayanan TN, Joy PA, Al-Omari IA, Sakthi Kumar D, Yoshida Y, Anantharaman MR. Template-assisted synthesis and characterization of passivated nickel nanoparticles. Nanoscale Research Letters. 2010; 5: 889–897.

[25] Kuppan B, Selvam P. Platinum-supported mesoporous carbon (Pt/CMK-3) as anodic catalyst for direct methanol fuel cell applications: the effect of preparation and deposition methods. Progress in Natural Science: Materials International. 2012; 22(6): 616–623.

[26] Wilkerson RJ, Elder T, Sowinksi O, Fostvedt JI, Hoefelmeyer JD. Phase transfer of oleic acid stabilized rod-shaped anatase TiO_2 nanocrystals. Surface Science. 2016; 648: 333–338.

[27] Langford JI, Wilson AJC. Scherrer after sixty years: a survey and some new results in the determination of crystallite size. Journal of Applied Crystallography. 1978; 11: 102–113.

[28] van Huis MA, van Veen A, Schut H, Eijt SWH, Kooi BJ, De Hosson JTM. Structural properties of Au and Ag nanoclusters embedded in MgO. Nuclear Instruments and Methods in Physics Research Section B. 2002; 191: 442–446.

[29] Biesinger MC, Lau LWM, Gerson AR, Smart RSC. Resolving surface chemical states in XPS analysis of first row transition metals, oxides and hydroxides: Sc, Ti, V, Cu and Zn. Applied Surface Science. 2010; 257: 887–898.

[30] Wöll C. The chemistry and physics of zinc oxide surfaces. Progress in Surface Science. 2007; 82: 55.

[31] Kónya Z, Vesselényi I, Kiss J, Farkas A, Oszkó A, Kiricsi I. XPS study of multiwall carbon nanotube synthesis on Ni-, V-, and Ni, V-ZSM-5 catalysts. Applied Catalysis A: General. 2004; 260(1):55–61.

[32] Loviat F, Czekaj I, Wambach J, Wokaun A. Nickel deposition on γ-Al_2O_3 model catalysts: an experimental and theoretical investigation. Surface Science. 2009; 603(14): 2210–2217.

[33] Contour JP, Mouvier G, Hoogewijs M, Leclere C. X-ray photoelectron spectroscopy and electron microscopy of PtRh gauzes used for catalytic oxidation of ammonia. Journal of Catalysis. 1977; 48: 217–228.

[34] Roth C, Goetz M, Fuess H. Synthesis and characterization of carbon-supported PteRueWOx catalysts by spectroscopy and diffraction methods. Journal of Applied Electrochemistry. 2001; 31: 793–798.

[35] Takasu Y, Unwin R, Tesche B, Bradshaw AM, Grunze M, Photoemission from palladium particle arrays on an amorphous silica substrate. Surface Science. 1978; 77: 219–232.

[36] Aricò AS, Antonucci V, Giordano N, Shukla AK, Ravikumar MK, Roy A, Barman SR, Sarma DD, Methanol oxidation on carbon-supported platinum–tin electrodes in sulfuric acid. Journal of Power Sources. 1994; 50: 295–309.

[37] Ledesma BC, Vallés VA, Rivoira LP, Martínez ML, Anunziata OA, Beltramone AR. Hydrogenation of tetralin over Ir catalysts supported on titania-modified SBA-16. Catalysis Letters. 2014; 144: 1–13.

[38] Sawatzky GA, Post D. X-ray photoelectron and Auger spectroscopy study of some vanadium oxides. Physical Review B. 1979; 20: 1546–1555.

[39] Sugano S, Tanabe Y, Kamimura H. Multiplets of transition-metal ions in crystals. Academic Press, New York (1970).

[40] Bassi AL, Cattaneo D, Russo V, Bottani C E, Barborini E, Mazza T, Piseri P, Milani P, Ernst FO, Wegner K, Pratsinis SEJ. Raman spectroscopy characterization of titania nanoparticles produced by flame pyrolysis: the influence of size and stoichiometry. Journal of Applied Physics. 2005; 98: 74305.

[41] Balaji S, Djaoued Y, Robichaud J. Phonon confinement studies in nanocrystalline anatase-TiO_2 thin films by micro Raman spectroscopy. Journal of Raman Spectroscopy. 2006; 37: 1416–1422.

[42] Mazza T, Barborini E, Piseri P, Milani P, Cattaneo D, Bassi AL, Bottani CE, Ducati C. Raman spectroscopy characterization of TiO_2 rutile nanocrystals. Physical Review B: Condensed Materials. 2007; 75: 045416.

[43] Konatowski J, Wichterlova B, Jitkovsky J, Loffler E, Pilz W. Spectroscopic studies of vanadium-substituted zeolitic silicates of MFI topology. Journal of the Chemical Society: Faraday Transactions. 1996; 92: 1067–1078.

[44] Yang Q, Wang S, Lu J, Xiong G, Feng ZC, Xin Q, Li C. Epoxidation of styrene on Si/Ti/SiO2 catalysts prepared by chemical grafting. Applied Catalysis A: General. 2000; 194: 507–514.

[45] Dresselhaus MS, Jorio A, Hofmann M, Dresselhaus G, Saito R. Perspectives on carbon nanotubes and graphene Raman spectroscopy. Nano Letters. 2010; 10: 751–758.

[46] Lu J, Yang J, Wang J, Lim A, Wang S, Loh KP. One-pot synthesis of fluorescent carbon nanoribbons, nanoparticles, and graphene by the exfoliation of graphite in ionic liquids. ACS Nano. 2009; 3: 2367–2375.

[47] Zhang W, Cui J, Tao C, Wu Y, Li Z, Ma L, Wen Y, Li G. A strategy for producing pure single-layer graphene sheets based on a confined self-assembly approach. Angewandte Chemie. 2009; 121: 5978–5982.

[48] Tuinstra F, Koenig JL. Raman spectrum of graphite. The Journal of Chemical Physics. 1970; 53: 1126–1130.

[49] Darmstadt H, Roy C, Kaliaguine S, Choi SJ, Ryoo R. Surface chemistry of ordered mesoporous carbons. Carbon. 2002; 40: 2673–2683.

[50] Zeinaldi F, Ghoreyshi AA, Najafpour GD. Adsorption of dichloromethane from aqueous phase using granular activated carbon: isotherm and breakthrough curve measurements. Middle East Journal Science Research. 2010; 5(4): 191–198.

[51] Kim B-J, Park S-J. Optimization of the pore structure of nickel/graphite hybrid materials for hydrogen storage. International Journal of Hydrogen Energy. 2011; 36: 648–653.

[52] Nguyen TQ, Bustria Padama AA, Sison Escano MC, Kasai, H. Theoretical study on the adsorption of NO on metal macrocycles, metal = Mn,Fe,Co,Ni,Cu,Zn. ECS Transactions. 2013; 45(20): 91–100.

[53] Psofogiannakisa GM, Froudakis GE. Fundamental studies and perceptions on the spillover mechanism for hydrogen storage. Chemical Communications. 2011; 47: 7933–7943.

[54] Hoang Tuan KA, Antonelli David M. Exploiting the Kubas interaction in the design of hydrogen storage materials. Advanced Materials. 2009; 21: 1787–1800.

[55] Shalabi AS, Taha HO, Soliman KA, Abeld S. Hydrogen storage reactions on titanium decorated carbon nanocones theoretical study. Journal of Power Sources. 2014; 271: 32–41.

[56] Zhang LP, Wu P, Sullivan MB. Hydrogen Adsorption on Rh, Ni, and Pd Functionalized single-walled boron nitride nanotubes. Journal of Physical Chemistry C. 2011; 115: 4289–4296.

Chapter 4

Mesoporous Carbons for Energy-Efficient Water Splitting to Produce Pure Hydrogen at Room Temperature

Mohindar S. Seehra and Vishal Narang

Additional information is available at the end of the chapter

http://dx.doi.org/10.5772/62503

Abstract

Theoretical and experimental aspects of the use of mesoporous carbons in carbon-assisted water electrolysis (CAWE) to produce pure hydrogen at room temperature are presented. It is shown that the electrical energy requirements for CAWE can be as low as 20% of the energy needed for conventional water electrolysis, the extra energy coming from the electrochemical oxidation of carbon occurring at room temperature. Although CO_2 is produced at the anode in this process, it is well separated from pure H_2 produced at the cathode. Experimental results are reviewed for a variety of carbons with the major focus on the results obtained with carbon BP2000, which has both mesopores and micropores and a nanocarbon produced by the hydrothermal treatment of microcrystalline cellulose.

Keywords: Hydrogen, Carbons, Electrolysis, Energy efficiency, Carbon dioxide

1. Introduction

Mesoporous materials with pore size from 2 to 50 nm fall in the middle range between microporous (pore size <2 nm) and macroporous (pore size >50 nm) materials. In a recent book edited by Titirici [1] and a review by Liang et al. [2], various processes for producing mesoporous carbon materials have been described. Sevilla and Fuertes [3] and Seehra et al. [4, 5] used hydrothermal treatment of cellulose and lignin to produce nanoparticles of carbon. Liu and Guo [6] compared the characteristics of carbons produced by the hydrothermal treatment of holocellulose and crude biomass. Other methods of preparing mesoporous carbons have been reported by Zhou et al. [7] for supercapacitor applications. Such carbons produced by different methods have applications in separation processes, catalysis, energy storage (such as super capacitors), and energy conversion. A very popular commercial source of mesoporous and microporous carbons is non-graphitized Black Pearls (BP) carbon blacks manufactured by Cabot Corporation. The pore structures of the BP carbons have been reported by Kruk et al. [8].

The focus of this review is on the use and the science of mesoporous/microporous carbons for the electrochemical production of pure hydrogen at room temperature employing electrical energy as low as 20% of the energy used in ordinary water electrolysis (WE). This process was first proposed and demonstrated by Coughlin and Farooque [9, 10] and developed more fully in the recent investigations of Seehra et al. using a variety of carbons and termed as carbon-assisted water electrolysis (CAWE) [11, 12]. Some follow-up studies of this process have been reported by Dubey et al. [13] using a carbon nanotube anode and by Giddy et al. [14] using a solid-state electrolytic cell. A different variation of this process involving hydrogen generation by laser irradiation of a carbon powder suspension in water has been reported by Akimoto et al. [15]. Results of the hydrogen evolution rate R_H using a variety of carbons show some proportionality to the surface area of a carbon which in turn is related to the pore structure of the carbons. The best results so far have been obtained with carbon BP2000, which as shown later has both micropores and mesopores and a surface area of about 1500 m^2/g.

This review covers the science behind CAWE followed by select experimental results and a comparative energy analysis of the process vis-à-vis conventional WE. Such comparative energy assessment of different processes to produce H_2 is essential for considering practical applications of the different processes. The remaining sections are organized as follows. In Section 2, prominent analytical techniques used in the structural characterization of porous carbons are reviewed with some examples imported from the published literature. Section 3 deals with the science behind CAWE and the experimental results on the electrochemical production of hydrogen reported in the literature using different carbons. The comparative energy analysis of the CAWE and WE processes is presented in Section 4. Concluding remarks of this review are given in Section 5.

2. Structural characterization of porous carbons

The porous structure of materials is often determined by means of adsorption of gases such as nitrogen and argon over a wide range of pressures. For activated carbons, the resulting structure depends on the process of activation, the activation agent, and the nature of the raw material used for producing a carbon. Kruk et al. [8] reported the surface and structural parameters (standard BET surface area S_{BET}, external surface area S_{ex}, micropore volume V_{mi}, and total pore volume V_t) of six carbon blacks produced by Cabot Corporation. These are named as Cabot BP120, BP280, BP460, BP800, BP1300, and BP2000 with S_{BET} = 30, 41, 78, 242, 520, and 1450 m^2/g, respectively. The pore volume distribution of these six carbons determined from nitrogen adsorption and desorption isotherms at 77.35 K and reproduced from Ref. [8]

http://dx.doi.org/10.5772/62503

is shown in **Figure 1**. From these graphs, it is evident that BP 120 and BP 280 have only small amount of pores, mostly in macroregion. BP460 has some mesoporosity, whereas BP800 and BP1300 are largely mesoporous with some microporosity. In contrast, BP2000 contains large fraction of micropores in addition to mesopores and some macropores. The ratio of S_{BET}/V_t for BP120, BP280, BP460, BP800, BP1300, and BP2000, based on the numbers given in Ref. [8] are 375, 410, 269, 356, 571, and 829 m^2/mL, respectively, showing that shapes and sizes of the particles and pores for the BP1300 and BP 2000 are quite different from the other four carbons.

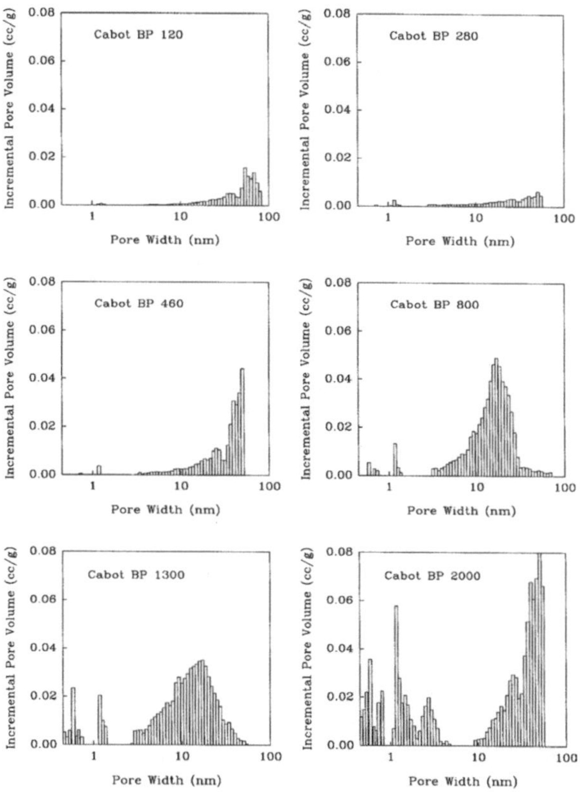

Figure 1. Pore size distribution of the six carbon blacks. (Reproduced with permission from Kruk et al. [8], copyright Elsevier 1996).

Characterization of seven activated carbons from PICA-USA for supercapacitor applications is reported by Gamby et al. [16]. Their BET surface areas ranged from a low of 1200 m^2/g to a high of $S_{BET} = 2315\ m^2/g$. These carbons contained roughly equal fractions of microporous and mesoporous volume fractions with total porous volume ≈1 cm^3/g for pore sizes <50 nm. The microporous volume increases nearly linearly with increase in the surface area. The specific capacitance increased by about 40% with increase in S_{BET} from about 1000 m^2/g to about 2000 m^2/g. This improvement in specific capacitance was tentatively related to increase in the mesoporous volume of the carbons.

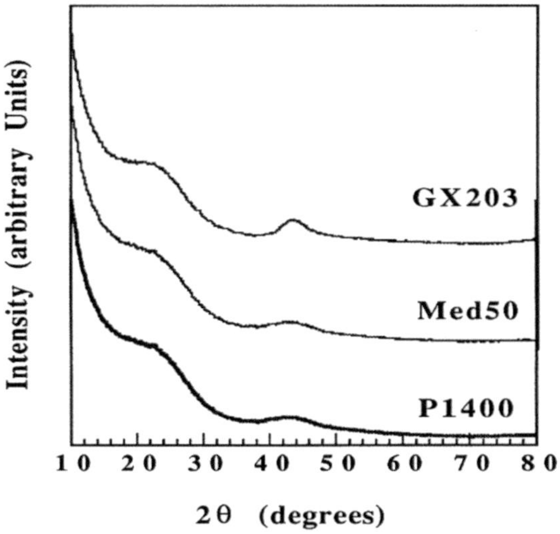

Figure 2. Wide-angle X-ray diffraction patterns of the three PICA carbons. (Reproduced with permission from Manivannan et al. [17], copyright Elsevier 1999).

Manivannan et al. [17] reported characterization of three PICA carbons viz. GX203, P1400, and Med50 with S_{BET} = 1000, 1150 and 2000 m^2/g, respectively, using the experimental techniques of wide-angle X-ray diffraction (WAXD), Raman spectroscopy, magnetometry, and electron spin resonance (ESR) spectroscopy. In **Figure 2**, we show the WAXD patterns of the three carbons using Cu-Kα source, with the broad lines near $2\theta = 24°$ and 44° due to graphitic crystallites corresponding to the (002) and (101)/(101) Bragg lines, respectively. The widths of these lines are then used to determine L_c (L_a) as the crystallite sizes along the c(a) directions, yielding $L_c \approx 1$ nm and $L_a \approx 3$ nm for these carbons. The Raman spectra of the three carbons, shown in **Figure 3**, can also be used to determine L_a from the empirical relation: L_a (nm) = 4.4/R, where R = I(1350)/I(1600) is the ratio of the intensities of the D band near 1350 cm^{-1} and the G band near 1600 cm^{-1}. This analysis also showed $L_a \approx 3$ nm in agreement with the results from WAXD. A similar comparison of the use of WAXD and Raman spectroscopy for characterizing carbons is given by Cuesta et al. [18]. ESR and magnetometry was used by Manivannan et al. [17] to determine the density of ESR active surface dangling bonds and nature of magnetic impurities which otherwise could not be detected by WAXD.

From the WAXD data shown in **Figure 2**, it is evident that for the lower 2θ values, the scattered intensity of X-ray photons increases sharply with decrease in 2θ. This is a characteristic feature of many amorphous materials. Measurements of the intensity of scattered X-rays covering smaller angles from $2\theta = 0.1°$ to 10° are termed small angle X-ray scattering (SAXS), and it can provide very useful information about pore sizes and structural aspects of macromolecules between 5 and 25 nm. In this case, the quantity plotted is the intensity I(q) of the scattered X-rays of wavelength λ as a function of momentum transfer $q = 4\pi \cdot \sin\theta/\lambda$ and such a plot can separate out contributions from micro and macropores [19]. In general, the intensity $I(q) \sim q^{-\alpha}$ where $\alpha = 6-D$, with D being the dimensionality of pore-boundary surface [20]. Experiments of SAXS in a lignite coal showed $\alpha = 3.5$ and so D = 2.5 signifying fractal nature of the surface.

Peaks in the I(q) versus q provides information on the size of macromolecules as for example observed in carbon nanotube suspensions [21].

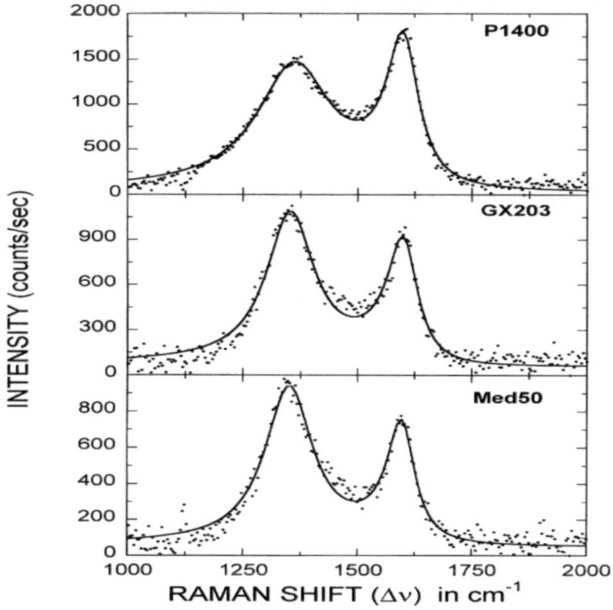

Figure 3. Raman spectra of the three PICA carbons. (Reproduced with permission from Manivannan et al. [17], copyright Elsevier 1999).

Techniques other than those described above that have also been used for the characterization of carbons include Fourier transform infrared (FTIR) spectroscopy, thermogravimetric analysis (TGA), X-ray photoelectron spectroscopy (XPS), scanning electron microscopy (SEM), and transmission electron microscopy (TEM). FTIR spectroscopy is particularly useful for determining the nature of surface functional groups. The use of all these techniques except XPS was employed in the characterization of the carbons produced from the hydrothermal carbonization of microcrystalline cellulose [4]. SEM/TEM is indispensable tools for visualization of the size and morphology of the particles [1–5], and TGA provides good information on the decomposition and oxidation characteristics of carbons [4]. The technique of XPS is often used for determining the elemental composition of a material. In summary, techniques that are used for structural characterization of other materials can also be used for carbons.

3. Carbon-assisted WE

The basic reactions for the well-known process of WE and Carbon-assisted water electrolysis (CAWE), the latter proposed by Coughlin and Farooque [9, 10], are given below:

3.1. Water electrolysis (WE)

$$\text{Anode: } 2H_2O \rightarrow O_2 + 4H^+ + 4e^- \tag{1}$$

$$\text{Cathode: } 4H^+ + 4e^- \rightarrow 2H_2 \tag{2}$$

$$\text{Net reaction: } 2H_2O \rightarrow 2H_2 + O_2 \tag{3}$$

3.2. Carbon-assisted water electrolysis (CAWE)

$$\text{Anode: } C(s) + 2H_2O \rightarrow CO_2 + 4H^+ + 4e^- \tag{4}$$

$$\text{Cathode: } 4H^+ + 4e^- \rightarrow 2H_2(g) \tag{5}$$

$$\text{Net reaction: } C(s) + 2H_2O(l) \rightarrow CO_2(g) + 2H_2(g) \tag{6}$$

To understand why CAWE is more energy efficient than WE, the change in the Gibbs-free energy $\Delta G = \Delta H - T\Delta S$ needs to be determined where ΔH (ΔS) is the change in enthalpy (entropy) of the system at temperature T. This is valid for the reactions under constant temperature and constant pressure conditions. Enthalpy is best understood as standard heat of formation of a compound from its basic elements at 25°C, the enthalpy being zero for the elements. For a compound to be stable, ΔH has to be negative. A reaction is favored if $\Delta G = \Delta G$ (products) − ΔG (reactants) < 0 which implies that a reaction is favored if, $\Delta H < 0$ and $\Delta S > 0$. We now apply these concepts to reactions for WE and CAWE listed in Eqs. (1)–(6).

At 25°C, ΔH (*l*) = −68.32 kcal/mol for water and ΔH (*g*) = −94.05 kcal/mol for CO_2 [22]. The standard molar entropies (in units of cal/mol K) are 31.2, 49.0, 16.7, and 51.0 for $H_2(g)$, $O_2(g)$, $H_2O(l)$, and $CO_2(g)$, respectively. Using these numbers for reaction (3) gives ΔH = 136.64 kcal and $T\Delta S$ = 23.23 kcal for 298 K, yielding $\Delta G = \Delta H - T\Delta S$ = 113.41 kcal > 0. ΔG being >0 means that the reaction (3) is not favored and energy must be supplied to split water into H_2 and O_2. Finally, ΔG is related to the electrical potential E^o generated in the reaction by the relation [22]:

$$\Delta G = -nFE^o \tag{7}$$

Here, n is the number of electrons involved in the reaction and F = 96484.56 C/mol is the Faraday constant. For WE, n = 4 according to Eq. (1) leading to E^o = −1.23 V. The negative sign implies that a minimum of E^o = 1.23 V must be applied to split water. Note that we have used 1 cal = 4.184 J and volt = Joules /Coulomb.

For CAWE, defined by reaction (6), a similar analysis yields ΔH = 42.59 kcal, ΔS = 80.05 cal leading to ΔG = 18.74 kcal and E^o = −0.20 V. This suggests that for reaction (6) to proceed, a minimum E^o = 0.20 V needs to be applied which is a factor of 1.23/0.20 $\simeq$ 6 smaller than that needed for reaction (3) of WE. This extra energy comes from carbon which gets oxidized to CO_2 in the process, making CAWE six times more energy efficient than WE, at least theoretically. This process of CAWE may thus be classified as electrochemical gasification of carbon at room temperature producing pure H_2 at the cathode which is well separated from pure CO_2 produced at the anode.

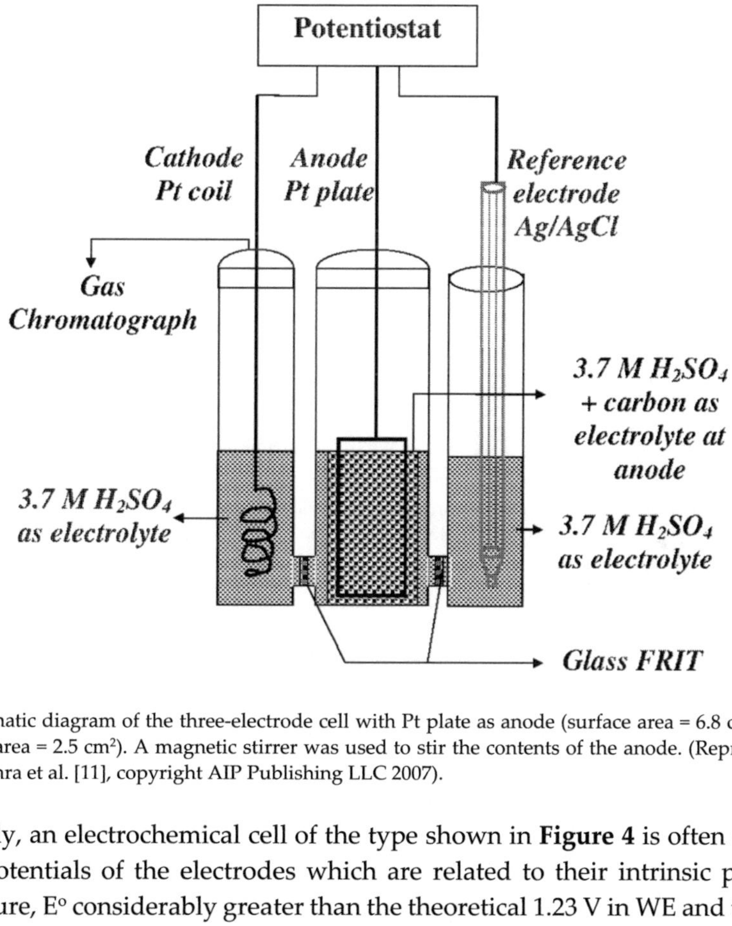

Figure 4. A schematic diagram of the three-electrode cell with Pt plate as anode (surface area = 6.8 cm^2) and Pt coil as cathode (surface area = 2.5 cm^2). A magnetic stirrer was used to stir the contents of the anode. (Reproduced with permission from Seehra et al. [11], copyright AIP Publishing LLC 2007).

Experimentally, an electrochemical cell of the type shown in **Figure 4** is often used. Because of the over-potentials of the electrodes which are related to their intrinsic properties and surface structure, E^o considerably greater than the theoretical 1.23 V in WE and the theoretical 0.20 V in CAWE needs to be applied to initiate H_2 evolution. Electrodes made of Pt and Pd are often used because they have the lowest over-potentials [23]. For each applied voltage E^o between the cathode and the anode, the quantities measured are as follows: (i) current I_o; (ii) the time t_H needed to produce the same amounts of H_2 for each E^o as measured by gas chromatography; and the computed quantities such as the hydrogen evolution rate $R_H = 1/t_H$ and the efficiency factor $A_H = R_H/E^oI_o$ representing evolved H_2 per kWh of energy used. The voltage E^o is with respect to the standard hydrogen electrode (SHE), and it is determined from the Eq.: E° = E(P) + 0.22 V where E(P) is the voltage measured by the potentiostat with respect to the reference electrode Ag/AgCl with E° = 0.22 V (SHE). It is noted that in many publications, voltages listed are simply E(P), and this needs to be kept in mind while comparing with the theoretical estimates.

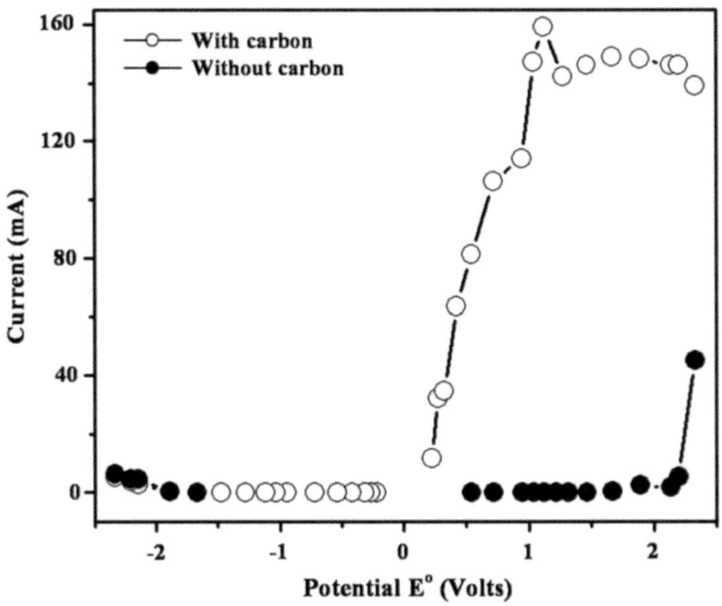

Figure 5. Forward and reverse current–voltage characteristics of the reaction, with and without adding carbon BP2000. (Reproduced with permission from Seehra and Bollineni [12], copyright Elsevier 2009).

In **Figure 5**, the effect of adding carbon BP2000 to the anode in the amount of 0.08 gm/cm^3 of the electrolyte (3.7 M H_2SO_4) is shown in terms of current–voltage plot. Without added carbon, an increase in current indicative of the start of WE only begins for E° > 2 V. With the addition of carbon BP2000 to the anode, a rapid increase in current accompanied by evolution of H_2 gas at the cathode was observed for E° > 0.4 V. These results are consistent with the theoretical predictions mentioned earlier in that the presence of carbon in the anode significantly lowers the threshold voltage to initiate WE and produce H_2 gas at the cathode.

It is noted that 3.7 M H_2SO_4 is chosen for the electrolyte since this particular concentration was found to have the highest electrical conductivity [10]. Similarly, the carbon concentration of 0.08 g/cm^3 was determined to be most efficient and practical by varying the concentration of the added carbon [11, 12]. A number of carbons were tried for their effects on the hydrogen generation rate R_H (**Figure 6**). As evident in **Figure 6**, carbon BP2000 yielded the best R_H values and so most of the follow up experiments were made with carbon BP2000 [12]. Another carbon that produced similar R_H values was the carbon produced by the hydrothermal treatment of microcrystalline cellulose using water as solvent and treated at 250°C for 60 min under the maximum pressure of 800 psi. This hydrothermal carbonization yielded spherical carbon particles of about 210 nm [4]. In the following section, the results obtained with carbon BP2000

http://dx.doi.org/10.5772/62503

[12] and hydrothermally treated cellulose are compared. The liquid-phase products obtained in the hydrothermal treatment of microcrystalline cellulose have also been analyzed [24].

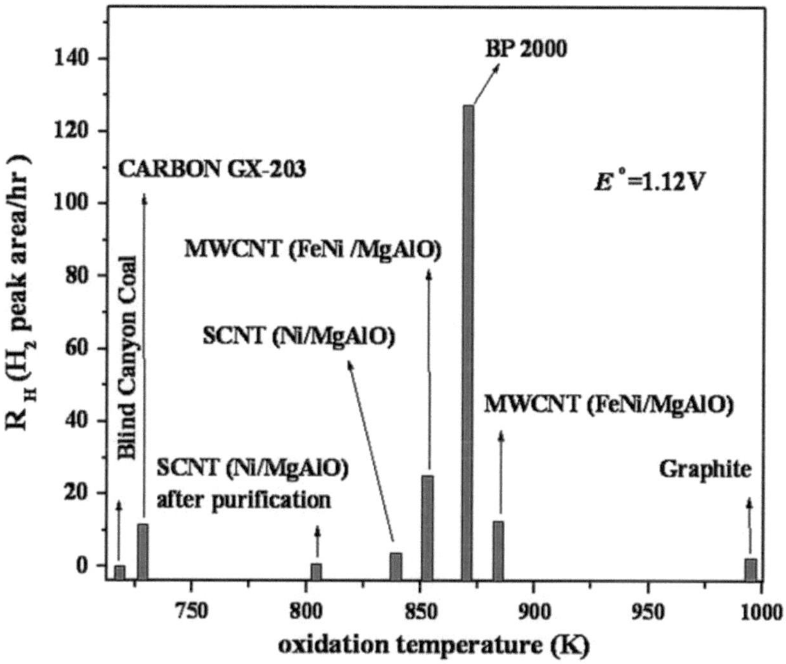

Figure 6. Comparison of hydrogen evolution rate R_H measured at E° = 1.12 V for several carbons plotted vs. oxidation temperatures of carbons determined by thermo-gravimetric analysis. MWCNT and SCNT stand for multi-walled and single-walled carbon nanotubes, respectively (Reproduced with permission from Seehra and Bollineni [12], copyright Elsevier 2009).

In the experiments reported in Refs. [4] and [12], H_2 was produced electrochemically first by conventional WE using 3.7 M H_2SO_4 as the electrolyte and then by adding carbons and cellulose-derived nanocarbons to the anode in separate experiments, repeating the experiments each time by measuring current I_o and hydrogen solution rate R_H as a function of the voltage E° applied between the cathode and anode. Evolved hydrogen at the cathode was measured by a calibrated gas chromatograph. The efficiency factor $A_H = R_H/E° I_o$ representing evolved H_2 per kWh of energy used is then computed for each E°. The plots of current in the circuit, R_H and A_H as a function of applied voltage E° (relative to the SHE) are shown in **Figure 7** for three cases: (i) electrolyte only, (ii) carbon BP2000 added to the anode in the amount of 0.08 gm/mL of the electrolyte, and (iii) the carbon produced by the hydrothermal treatment of cellulose added to the anode in the amount of 0.08 gm/mL of the electrolyte. In the WE process, a noticeable current and R_H is observed only if E° > 2 V. However, for both BP2000 and hydrothermally treated cellulose, significant I_o and R_H are observed at E° = 0.7 V. In **Figure 7**, the horizontal dotted line is drawn to show that using commercial carbon BP2000 with surface area of 1500 m^2/g, same R_H is obtained at applied voltage E° = 0.7 V using carbon BP2000 as obtained by applying E° = 2.6 V in conventional WE. For carbon produced by hydrothermal treatment of cellulose, E° = 1.2 V is needed to produce the same R_H. A detailed analysis of the comparison of the efficiencies of the three processes is given in Section 4.

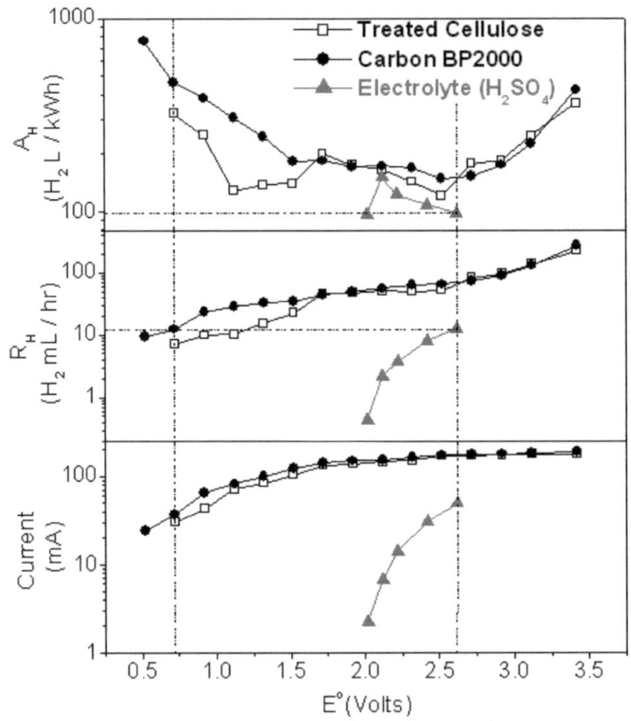

Figure 7. Comparison of current, hydrogen evolution rate R_H, and efficiency factor A_H against applied voltage E° (vs. standard hydrogen electrode) for HTP-cellulose treated at 250°C for 60 min, carbon BP2000 and electrolyte only. (Reproduced with permission from Seehra et al. [4], copyright Elsevier 2012).

According to the Eqs. (4) and (5) valid for CAWE, H_2 should be produced at the cathode and CO_2 at the anode in the molar ratio of H_2/CO_2 =2. With careful quantitative measurements, this indeed was found to be valid using carbon BP2000 [12]. However, using carbon produced from the hydrothermal treatment of cellulose, CO_2 at the anode could not be detected with certainty suggesting that the reaction with this carbon is more complicated perhaps because of the presence of surface group present in this carbon [4]. Another difficulty in the quantification of CO_2 is that its peak in gas chromatography is quite weak relative to peaks from H_2 and O_2. Regardless, CO_2 produced in CAWE at the anode is well separated from H_2 produced at the cathode. This is important since CO_2 is a green-house gas and so it can be collected in relatively pure form and so appropriately used or sequestered.

Of all the carbons results for which are shown in **Figure 6**, BP2000 was found to be the most efficient for producing H_2. It is very likely that this is due to surface area effect since BP2000 has high surface area of 1500 m^2/g. However, a direct correlation between the surface area and R_H has not yet been established experimentally. It may be possible to do so by comparing R_H yields for the various BP carbons with different surface areas shown in **Figure 1**.

4. Comparative energy analysis

Understanding the efficiencies of various processes to produce hydrogen is very important for practical considerations. In this regard, Rosen and Scott [25] have published a comparative energy analysis of the various processes currently in use for producing H_2. The summary of this analysis, reproduced from their paper, is given in **Table 1**. According to this analysis, the energy efficiency of the conventional WE for producing H_2 gas is only about 30% compared to about 85% for the steam methane reforming (SMR) and about 60% for coal gasification (CG) processes. If heat losses to the environment are also included (exergy) in the analysis, then these efficiencies are reduced by an additional few percent. The energy efficiency here is defined as the ratio of the energy contained in the produced hydrogen to the input energy. Because of the higher efficiencies of the SMR and CG processes, these are currently the favored techniques for producing H_2 although both of these processes also produce CO_2.

Category	**Process**	**Efficiency (%)**	
		Energy	**Exergy**
Hydrocarbon-based	Steam-methane reforming (SMR)	86	78
	Coal gasification	59	49
Non-hydrocarbon-based	Current-technology water electrolysis	30	26
	Advanced-technology water electrolysis	49	41
	Thermochemical water decomposition	21	19
Integrated	SMR/current-technology water electrolysis	55	48
	SMR/advanced-technology water electrolysis	70	62
	SMR/thermochemical water decomposition	45	40

Reproduced with permission from Rosen and Scott [25] copyright Elsevier 1998

Table 1. Hydrogen production processes and their efficiencies considering fuel and/or a hypothetical heat source as the external energy inputs.

In the recent paper [4], energy efficiencies of the WE and CAWE processes were considered in terms of the energy equivalence of electrical energy in kWh (kilo-watt-hour) and liters (L) of produced H_2. Using the data available in the literature, it can be shown that one gasoline gallon equivalent (GGE) equals 33.4 kWh of electricity which in turn equals 10.1 m^3 of H_2 gas at STP (standard temperature and pressure). According to this equivalence, one kWh of electricity equals 300 L (liters) of hydrogen implying that one kWh of electricity used in electrolysis must produce 300 L of H_2 @ STP for 100% energy efficiency. This equivalence is used here to determine the energy efficiencies of WE and CAWE processes, the latter using carbon BP2000 and the carbon produced from the hydrothermal treatment of microcrystalline cellulose [4].

Examining the results shown in **Figure 7**, it is evident that for electrolyte only representing conventional WE, $E^o > 2$ V is needed to start the reaction, reaching close to saturation at $E^o =$

2.6 V. At this voltage, R_H is about 15 mL/h of hydrogen and A_H equals 100 L of H_2 produced per kWh of electrical energy used. As noted earlier, if process was 100% energy efficient, A_H = 300 L/kWh should have been produced. Thus, conventional WE is only about 33% (100/300) efficient, in close agreements with the results from the analysis of Rosen and Scott shown in **Table 1**.

Following the horizontal dotted line drawn in **Figure 7**, it is evident that using commercial carbon BP2000 with surface area of 1500 m^2/g, same R_H is obtained at applied voltage $E^o = 0.7$ V using carbon BP2000 as obtained by applying $E^o = 2.6$ V in conventional WE. However, at $E^o = 0.7$ V with carbon BP2000, $A_H = 450$ L/kWh of H_2 is produced resulting in energy efficiency of 150% (450/300) compared to about 33% for conventional WE. This extra energy representing nearly fivefold improvement in energy efficiency over conventional WE is coming from the extra electrons provided by use of carbon BP2000. This electrochemical gasification occurring at room temperature does produce CO_2 at the anode, but it is well separated from the H_2 produced at the cathode.

In **Figure 7**, the data of H_2 production using carbon produced from hydrothermally treated cellulose instead of carbon BP2000 is also shown. With the use of carbon produced from hydrothermally treated cellulose, again following the horizontal line, somewhat lower $A_H \simeq$ 350 L/kWh is obtained at $E^o = 0.7$ V with essentially similar evolution rate R_H of H_2. This yields slightly lower energy efficiency of about 120% (350/300) which is still a factor of four improvement over conventional WE. This electrochemical process with cellulose-derived nanocarbon has the added advantage that no CO_2 could be detected at the anode presumably because cellulose-derived nanocarbons have surface functional groups [4] unlike carbon BP2000. It is noted that in doing these comparisons of energy efficiencies, the cost of producing the carbons is not included. However, BP2000 is available commercially in large quantities and hydrothermal carbonization is environmentally friendly process employing water as solvent under subcritical temperature–pressure conditions [24].

In summary, the comparative energy analysis described here shows factors of four to five improvements in the energy efficiencies for producing H_2 electrochemically using commercial carbon BP2000 and cellulose-derived nanocarbons. Pure CO_2 is produced with the use of carbon BP2000, but it is well separated from H_2. The use of post-HTP cellulose has the advantage that no CO_2 could be detected in the process. For more practical issues related to the electrochemical production of hydrogen from WE, the reader is referred to the recent review by Wang et al. [23].

5. Concluding remarks

In this review, the focus has been on just one application of mesoporous/microporous carbons viz. in CAWE to produce H_2 at energy-efficient voltages. Theoretical considerations and experimental results have established that the electrical energy requirements for CAWE are a factor of about five smaller than those needed for conventional WE. CO_2 is produced at the cathode in CAWE resulting from the room temperature oxidation of carbon. However, it is well separated from pure H_2 produced at the anode. Whether this energy advantage of CAWE vis-à-vis conventional WE can be commercially exploited still remain to be seen since there are often other technical issues that must be overcome as described in the recent review of Wang et al. [23]. Comparing the results in **Figure 6** obtained with the use of a variety of carbons, it is evident that the best H_2 evolution rate R_H was obtained using carbon BP2000 with high surface area of about 1500 m^2/g. Since smaller pore size leads to higher surface area, a correlation quite likely exists between R_H and pore size of the carbons. Additional studies along these lines using different carbons with known pore-size structure and surface areas might be fruitful. As mentioned in the Introduction, there are of course many other applications of mesoporous and microporous carbons and some of these are addressed in the other chapters of this book.

Author details

Mohindar S. Seehra* and Vishal Narang

*Address all correspondence to: mseehra@wvu.edu

Department of Physics and Astronomy, West Virginia University, Morgantown, USA

References

[1] Titirici M.-M., *Sustainable carbon materials from hydrothermal processes*. Wiley Online Library, 2013.

[2] Liang C., Li Z., and Dai S., *Mesoporous carbon materials: synthesis and modification*. Angewandte Chemie International Edition, 2008. 47(20): 3696–3717.

[3] Sevilla M., and Fuertes A.B., *The production of carbon materials by hydrothermal carbonization of cellulose*. Carbon, 2009. 47(9): 2281–2289.

[4] Seehra M., Akkineni L., Yalamanchi M., Singh V., and Poston J., *Structural characteristics of nanoparticles produced by hydrothermal pretreatment of cellulose and their applications for electrochemical hydrogen generation*. International Journal of Hydrogen Energy, 2012. 37(12): 9514–9523.

[5] Seehra M.S., Pyapalli S.K., Poston J., Atta-Obeng E., and Dawson-Andoh B., *Hydrothermal conversion of commercial lignin to carbonaceous materials*. Journal of the Indian Academy of Wood Science, 2015. 12(1): 29–36.

[6] Liu F., and Guo M., *Comparison of the characteristics of hydrothermal carbons derived from holocellulose and crude biomass*. Journal of Material Science, 2015. 50: 1624–1631.

[7] Zhou D.-D., Du Y.-J., Song Y.-F., Wang Y.-G., Wang C.-X., and Xia Y.-Y., *Ordered hierarchical mesoporous/microporous carbon with optimized pore structure for supercapacitors*. Journal of Materials Chemistry A, 2013. 1(4): 1192–1200.

[8] Kruk M., Jaroniec M., and Bereznitski Y., *Adsorption Study of Porous Structure Development in Carbon Blacks*. Journal of Colloid and Interface Science, 1996. 182(1): 282–288.

[9] Coughlin R.W., and Farooque M., *Hydrogen production from coal, water and electrons*. Nature, 1979. 279(5711): 301–303.

[10] Coughlin R.W., and Farooque M., *Electrochemical gasification of coal-simultaneous production of hydrogen and carbon dioxide by a single reaction involving coal, water, and electrons*. Industrial & Engineering Chemistry Process Design and Development, 1980. 19(2): 211–219.

[11] Seehra M., Ranganathan S., and Manivannan A., *Carbon-assisted water electrolysis: an energy-efficient process to produce pure H_2 at room temperature*. Applied Physics Letters, 2007. 90(4): 044104; ibid 2008, 92, 239902.

[12] Seehra M., and Bollineni S., *Nanocarbon boosts energy-efficient hydrogen production in carbon-assisted water electrolysis*. International Journal of Hydrogen Energy, 2009. 34(15): 6078–6084.

[13] Dubey P.K., Sinha A.S.K., Talapatra S., Koratkar N., Ajayan P.M., and Srivastava O.N., *Hydogen generation by water electrolysis using carbon nanotube anode*, International Journal of Hydrogen Energy, 2010. 35: 3945–3950.

[14] Giddy S., Kulkarni A., and Badwal S. P. S., *Low emission hydrogen generation through carbon assisted electrolysis*, International Journal of Hydrogen Energy, 2015. 40: 70–74.

[15] Akimoto I., Maeda K., and Ozaki N., *Hydrogen generation by laser irradiation of carbon powder in water*, Journal of Physical Chemistry C, 2013. 117: 18281–18285.

[16] Gamby J., Taberna P.L., Simon P., Fauvarque J.F., and Chesneau M., *Studies and characterisations of various activated carbons used for carbon/carbon supercapacitors*. Journal of Power Sources, 2001. 101(1): 109–116.

[17] Manivannan A., Chirila M., Giles N., and Seehra M., *Microstructure, dangling bonds and impurities in activated carbons*. Carbon, 1999. 37(11): 1741–1747.

[18] Cuesta A., Dhamelincourt P., Laureyns J., Martinez-Alonso A., and Tascon J.M., *Comparative performance of X-ray diffraction and Raman microprobe techniques for the study of carbon materials*. Journal of Materials Chemistry, 1998. 8(12): 2875–2879.

[19] Gibaud A., Xue J., and Dahn J., *A small angle X-ray scattering study of carbons made from pyrolyzed sugar*. Carbon, 1996. 34(4): 499–503.

[20] Bale H.D., and Schmidt P.W., *Small-angle X-ray-scattering investigation of submicroscopic porosity with fractal properties*. Physical Review Letters, 1984. 53(6): 596–599.

http://dx.doi.org/10.5772/62503

[21] Zhou W., Islam M., Wang H., Ho D., Yodh A., Winey K.I., and Fischer J.E., *Small angle neutron scattering from single-wall carbon nanotube suspensions: evidence for isolated rigid rods and rod networks*. Chemical Physics Letters, 2004. 384: 185–189.

[22] Roberts E.K., *Principles of Physical Chemistry*. Allyn & Bacon, Boston, USA, 1984.

[23] Wang M., Wang Z., Gong X., and Guo Z., *The intensification technologies to water electrolysis for hydrogen production–a review*. Renewable and Sustainable Energy Reviews, 2014. 29: 573–588.

[24] Seehra M., Popp B., Goulay F., Pyapalli S., Gullion T., and Poston J., *Hydrothermal treatment of microcrystalline cellulose under mild conditions: characterization of solid and liquid-phase products*. Cellulose, 2014. 21(6): 4483–4495.

[25] Rosen M.A., and Scott D.S., *Comparative efficiency assessments for a range of hydrogen production processes*. International Journal of Hydrogen Energy, 1998. 23: 653–659.

Chapter 5

Biomass, Abundant Resources for Synthesis of Mesoporous Silica Material

Adebola Iyabode Akinjokun,
Tunde Victor Ojumu and
Aderemi Okunola Ogunfowokan

Additional information is available at the end of the chapter

http://dx.doi.org/10.5772/63463

Abstract

The synthesis of mesoporous silica materials was reviewed with a view to discuss the reaction mechanism and the various attempts made at enhancing the materials' properties by utilizing varieties of templating agents and silica frameworks from pure synthetic chemicals. This chapter also reviewed studies in which either the template or the framework was synthesized from benign reagents obtained from renewable sources, to achieve enhanced material properties. The view was to encourage the development of mesoporous silica materials in which both the template and the silica framework are from biomass origin. This approach may promote the large-scales synthesis of mesoporous silica for commercial purposes, which had previously been hampered by the toxic nature, cost of synthetic chemical reagents, and unsustainable synthetic routes.

Keywords: mesoporous silica materials (MSiM), templating agents, biomass, silica framework, micelle concentration, surfactant

1. Introduction

The facile discovery of M4IS family of mesoporous silica material is one of the significant breakthroughs in material science in the last two decades. This material with pore diameter between 2 and 50 nm has continued to be a subject of research focus till date, as its potential application in areas such as medicine, catalysis, adsorption, sensing, and photonics continues to rise. This is shown by more than 19,000 literature citations since 1993 along with numerous patents on the subject matter annually. However, the large-scale commercialization of the numerous research findings and patents in these fields has not been feasible. This may be due to the high cost of reagents (surfactants and silica precursor) combined with environmental concerns occasioned by the high toxicity of majority of the surfactants and precursors system used in the synthesis. Thus, the large-scale synthesis of cheap and environmentally safe mesoporous silica material is one of the greatest challenges in modern material syntheses.

Recently, there have been renewed interests in mesoporous silica-based material synthesis fueled by the new synthetic approach which lay emphasis on the use of biologically derived materials found in nature or extracted from biomass resources. It is generally believed that the cost and the ecological impacts of these syntheses would be drastically reduced, thus increasing the possibility of commercialization [1] in the nearest future. In addition, these biosynthesized materials often have well-defined hierarchical structures/network and properties not hitherto seen in those synthesized via utilization of synthetic chemical reagents. Therefore, this chapter provides a critical assessment of the synthetic approach used in the synthesis of mesoporous silica materials with a view to encourage biomass utilization as potential feedstock for the synthesis if scale-up development of mesoporous silica materials is envisioned at reasonable cost and with relatively less impact on the environment.

The review is organized into four sections. Section 2 reviews the background into the synthesis of mesoporous silica material (MSiM) from commercially sourced reagents (surfactants and silica framework) with a view to provide insight into the effects of choice of reagents on the structure and characteristics of the synthesized materials. While Section 3 focuses on the progress made in the part utilization of biological/biomass materials and the effects on their inclusion on pore size, surface area, and morphology in comparison with materials derived from purely synthetic chemicals. Finally, Section 4 emphasizes on the view to maximize the potential of biomass waste materials in advanced material synthesis.

2. Mesoporous silica synthesized by liquid crystal templating approach

2.1. Background into the synthesis of MSiM from conventional reagents

Mesoporous materials, an upper pore extension of microporous zeolithic materials, have attracted much attention due their large surface area (>1000 m^2/g), pore volume (app. 1 g/cm^3), and narrow pore distribution that can be adjusted between 2 and 50 nm to meet specific needs and functions. The commercial interest in their applications in catalysis, photonics, sensing, chiral separation, and drug delivery are additional incentives to focus on their synthesis. Pioneered by the facile discovery of the M4IS family (including MCM-41, MCM-48, and MCM-50) of mesoporous materials in 1992 [2]. The synthesis of MCM-41, the first and most popular member of the M4IS family, was facilitated by assembly of long chain cationic surfactant into micelles or lyotropic liquid crystals (LCC) at a specific surfactant concentration in aqueous solution known as *critical micelle concentration* (CMC). Hexadecyltrimethylammonium ion, a cationic surfactant, was used as the template-forming agent. These amphiphilic molecules assembled in polar solvents forming a hexagonal array of cylindrical micelles in which the non-polar hydrophobic water-repelling tail gathered in the centre and the polar hydrophilic water-loving head on the surface. This micelle served as template or pore-forming agent for the subsequent sol-gel hydrolytic polycondensation of inorganic silicate precursor, an alkoxysilanes (such as tetraethyl orthosilicate) under hydrothermal conditions in solution [3], forming an organic-inorganic composite. When the composite is calcined, the template is removed leaving the porous network. But earlier in 1990, Yanagisawa et al. [4] obtained a material which was later designated as folder sheet mesoporous materials (FSM)-16 by intercalating kanemite, a layered silicate with organic alkyltrimethylammonium chloride, a cationic surfactant. During the interaction, the SiO_2 layers in the organic-kanemite complexes were condensed to form 3-D SiO_2 networks which upon calcination yielded products having pores diameter 2–4 nm and large surface area (ca. 900 m^2/g). However, at the time of the publication, the materials were thought to be microporous as a mixed phase of surfactant and precursor was obtained.

In the report of Beck et al. [2], self-assembly and electrostatic charge matching between the surfactant and inorganic precursors was believed to be the driving force for the formation of mesostructure. Later, Tanev and Pinnavaia [5] reported a new approach to the synthesis of mesoporous materials via a neutral surfactants-templating approach which complemented the ionic-dependent pathways. This approach is based on hydrogen bonding and self-assembly between neutral amines and neutral inorganic precursor at ambient temperature. Due to weaker repulsive interaction between the neutral surfactant and precursor at the surfactant-solution interface, structures with thicker framework walls, small scattering sizes, and textural mesoporosities matching with those of materials produced from ionic pathways were obtained. The thicker walls supplied the material with improved thermal and hydrothermal stability and also made the environmentally benign recovery of the template via solvent extraction possible. In addition, the most significant chemical advantage of this approach perhaps is the synthesis of materials that cannot ordinarily be assessed by ionic templating approaches (e.g., alumina). In spite of the obvious advantages of the neutral templating route over the ionic route, the neutral surfactants still suffer from the same negative aspect as the cationic quaternary ammonium surfactants such as high cost and toxicity are therefore not suitable for large-scale syntheses of mesoporous materials. The original neutral templating approach was extended to obtained phases with increased wall thickness and thermal stability by the use of block copolymers (BCs). BCs such as the polyethylene oxide (PEO) polymeric surfactants were used for the synthesis of disordered worm-like phases at neutral pH and triblock copolymers under acidic condition to obtain highly ordered 2-D Santa Barbara Amorphous (SBA). However, despite the low cost of BCs, the surfactant showed poor degradability and toxicity to organisms, thereby hindering its utilization in large-scale syntheses.

2.2. Synthetic pathways

Several investigations into the synthesis of mesoporous silica materials have concluded that two different mechanisms are involved: (a) On the one hand, it involved the self-assembly of the surfactant (at high concentration) into a lyotropic crystalline phase (micelles) independent of the crystallization and subsequent polymerization of the silica framework around the preformed micellar aggregates framework. (b) On the other hand, it is also possible to produce this phase at lower surfactant concentration; the cooperative self-assembly of the surfactant and inorganic species is directed by the aqueous inorganic silica precursor to obtain liquid crystalline phase with hexagonal, cubic, or lamellar shape (**Figure 1**)

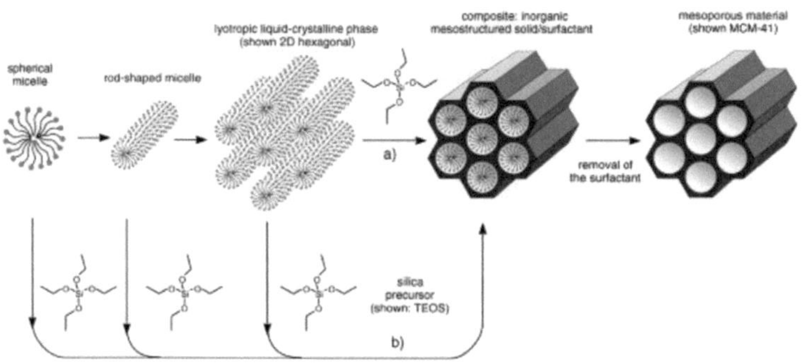

Figure 1. Formation of mesoporous materials by surfactant structure directing agents: (a) true liquid-crystal template mechanism, (b) cooperative liquid-crystal template mechanism [2].

A basic condition for the mesophase formation is that there must be an attractive interaction between the template and silica precursor which guarantees the subsequent inclusion of the structure-directing agent (SDA) in the final material without the occurrence of phase separation. According to Huo et al. [6] these interactions can be classified as follows: (a) under basic conditions, where the silica species are present as anions and cationic quaternary ammonium surfactants are used as the SDA, the synthetic pathway is termed (S^+I^-). (b) The original synthesis of MCM-41 is based on this pathway. In the charged reversed phase, under acidic conditions (i.e., below the isoelectronic point of the silica species; pH = 2) where the silica species are positively charged, an anionic SDA (S^-) is used to direct the self-assembly of cationic inorganic silica species (I^+) through the S^-I^+ pathway. (c) To produce an interaction with the cationic surfactant, a mediator ion X^- (usually halides) may be added to direct the mesophase formation via the $S^+X^-I^-$. (d) Conversely, when negatively charged surfactants (S^-) are used, it is also possible to work under basic condition whereby a mediator ion M^+ (usually alkali ions) can be added to ensure interaction between equally negatively charged species via $S^-M^+I^-$ pathway. Therefore, the interactions in pathways (a)–(d) are predominantly electrostatic in nature. In addition, it is possible for the attractive interactions to be mediated by hydrogen bonds; this is the case when neutral (S^0, e.g., long-chained amine) or nonionic surfactants (N^0, e.g., polyethylene oxide) and uncharged silica species (S^0I^0 or N^0I^0) or ion pairs (S^0H^+) (X^-I^0) pathways are present.

Depending on the nature, concentration, and temperature, surfactants in lyotropic systems can form different liquid crystal phases which includes isotropic micellar, cubic micellar, lamellar, or reversed-phase micellar phases. Therefore, the behavior of surfactants is the key for the controlled synthesis of silica mesostructure. As proposed by Israelachvili et al. [7], the relative

http://dx.doi.org/10.5772/63463

stabilities of the different micellar aggregate shapes and the mesophase structures can be predicted from a combination of molecular packing considerations and general thermodynamic principles. He further states that the preferred shape of the self-assembled surfactant molecules above its CMC depends on the effective mean molecular parameters of the micelle that establish the value of a dimensionless packing parameter g, which is defined as $g = V/a_o l_c$, where V is the effective volume of the hydrophobic chain, a_o is the mean aggregate surface area per hydrophilic head group, and l_c is the critical hydrophobic chain length [7]. The parameter g depends on the molecular geometry of the surfactant molecules, such as the number of carbon atoms in the hydrophobic chain, the degree of chain saturation, and the size and charge of the polar head group. In addition, the effects of solution conditions including ionic strength, pH, co-surfactant concentration, and temperature are included in V, a_o, and l_c.

3. Utilization of materials from renewable biomass as alternative reagents

Different processing parameters such as the type of surfactant and precursor among others determine the porosity and morphology of the final material. Mesoporous materials have been largely synthesized from synthetic/commercially sourced surfactants and precursors. However, the high cost, toxicity, and environmental impacts of these reactants and syntheses have led to the search for alternative functional replacements and environmentally benign synthetic routes. Recently, biomass, a renewable resource, has been identified as an abundant source of diverse surfactant materials such as polysaccharides, polypeptides and polyaromatics, and silica frameworks (from rice husk, wheat husk, etc.) that could be applied in silica mineralization. Many biomass materials exhibit hierarchical large-scale order and self-assembly properties that can be replicated in a variety of structures; hence, the use of biomass materials as alternative to commercial surfactants and precursors not only yields materials with structural diversities but also holds promise for the synthesis of low-cost completely biocompatible materials with reduced toxicity and application in areas of human medicine.

3.1. Utilization of surfactants derived from biomass resources

In the last few years, several efforts have been made in the production of surfactants with low toxicity and biodegradability; this includes the use of surfactant building blocks (hydrophilic/polar head and hydrophobic tail) derived from renewable resource and the use of surfactants derived from renewable polymer- and colloid-based biomacromolecules such as cellulose and chitosan. With respect to renewable hydrophobic segment of the surfactant, natural fats and oils, which are composed of wide distribution of fatty acids, have been applied in the synthesis of mixtures of surfactants. One of the limitations associated with the use of templates/SDA derived in material synthesis is that they are often obtained in mixtures containing anionic surfactants. The challenge here is that each of the components may interact differently with the silica precursor preventing a suitable organization of the porous network. Moreover, their purification may lead to increase in cost as well as adverse environmental impacts. However, the advantage inherent in this approach is the abundant availability of this sources and their fast biodegradability.

Canlas and Pinnavaia [8] reported the utilization of naturally derived oleyl amine surfactants for the synthesis of worm-like and lamellar mesoporous silica. These structures are similar to those obtained using petroleum-based alkyl amine surfactant via the S^0I^0 templating mechanism thus showing that surfactants from renewable source could effectively be used as templates in mesoporous silica synthesis. However, at lower solvent polarity, the surfactant, due to the flexibility of the hydrophobic chain, adopted a hairpin and other extended structures generating mesoporous silica with pore size in the supermicroporous range, thus making them more flexible as templating agents than the alkyl amines. This supermicroporous derivative is believed to offer greater size and shape selection separation and catalysis in comparison to their mesoporous analogues. Oleyl amine is obtained by the transformation of oleic acid derived from tallow fats or vegetable oils such as soya or palm oil. With respect to utilization of hydrophilic segment from renewable resource in generating eco-friendly surfactants, amino acid derived surfactant from another interesting group of renewable surfactants known for their excellent surface activity, biocompatibility, and low toxicity. Thomas et al. [9] reported the use of surfactant formulations based on glutamic acid and leucine containing a mixture of lauronyl amino acid and fatty acid as SDA for the formation of mixtures of ordered mesoporous silica including lamellar phases. This approach was challenging as the synthesis was carried out at neutral pH and room temperature using organic and inorganic reactants from renewable sources. Measurement of amino acid-based SDA CMC, surface tension at CMC (ST), foaming capacity, and foaming stability showed that the surfactant mixtures with the longest chain have properties comparable to petroleum-based surfactants [10]. Sodium N-dodecyl glycine ($C_{12}H_{25}NHCH_2COONa$) and potassium N-dodecyl glycine ($C_{12}H_{25}NHCH_2COOK$) amino acid-based surfactants prepared by reacting coconut oil amine with monochloroacetic acid in alkali medium were used to synthesize mesostructures different from lamellar, more precisely the synthesis of hexagonal and Pm3n cubic structures [11].

Another interesting class of surfactant with polar groups derived from renewable resources is the alkyl glycosides with growing interest due to their liquid crystalline phase-forming ability, chiral properties, non-toxicity, and biodegradability. Alkyl glycosides are another class of surfactants with polar head derived from renewable resources. The development of well-ordered materials using this sugar-based nonionic surfactant is however limited because the interaction between the weakly charged inorganic silica precursor and the nonionic surfactant are short-ranged hydrogen bonding often resulting in disordered mesostructure. The deposition of mesostructure silica films via the self-assembly of commercially available sugar-based alkyl glycosidic nonionic surfactants, such as n-octyl β-D-glycopyranoside (β-C_8Glc), n-dodecyl β-D-maltopyranoside (β-C_{12}Mal), and n-decyl β-D-maltopyranoside(β-C_{10}Mal), in combination with a prehydrolysed silicate, and the effects of size of sugar head group and alkyl chain length on the type of supramolecular assembly of the surfactant have been reported [12]. Although the three alkyl glycoside surfactants were successfully utilized in the sol-gel processing of crack-free mesoporous silica films, the hydrogen bonding interactions with the silicate species during the gel formation suppressed the siloxane condensation. β-C_{10}Mal and β-C_{12}Mal with bigger maltose head group formed curved mesophases at lower concentrations and lamellar phases at higher concentrations that were stable to template removal by solvent extraction or calcination. β-C_{12}Mal-templated and β-C_{12}Mal-calcined mesostructure silica film exhibited a significantly higher d spacing (41 Å) than the corresponding β-C_{10}Mal-templated silica (30 Å).

Renewable biomass wastes generated by agro-processing and food industries, including polymers and polymer colloids belonging to the polysaccharide family, constitute a potential field of research for obtaining new sustainable surfactants for the design of functional mesostructure. This is due to their availability, non-competitive use in the food industry, and more importantly, the rich chemical functionality and easy formulations [13]. Various polysaccharide-based polymers have been used as templates to direct the synthesis of silica; these include chitosan and chitin colloids and cellulose. Chitin, which is found in the exoskeletons of crustaceans, mollusk, and insects in nature, is the second most abundant renewable polysaccharide in nature. However, its use in design of functional material is limited by its poor solubility, but on partial deacetylation, it yields chitosan, a water-soluble substance furnished with several free amino acid groups. Due to its solubility in acidic media and its rich chemical groups, chitosan has been used for structuring silica-based materials [13–15]. However, these efforts yielded poorly ordered materials after template removal by calcination.

Recently, Alonso and Belamie [16] presented a versatile and novel approach involving the use of chitin colloids ($L = 260 \pm 80$ nm, $D = 23 \pm 3$ nm) from shrimps which are made up of bundles of chitin nanorod ($D = 3.2 \pm 0.6$ nm) with amino groups on their surface. The mechanism of formation is governed by the chitin nanorods self-assembly coupled with chitin-siloxane soft attractive attractions, which bore similarities with cooperative and dynamic template mechanisms proposed earlier [17, 18]. Nguyen et al. [19] reported the ordering of mesoporous silica with high surface area and layered structure templated by chitin nanocrystals (ChNCs) derived from king crab shell waste. ChNCs are spindle-shaped aqueous suspension of chitin nanorod prepared by sequential deacetylation and hydrolysis of chitin [20]. The ChNCs self-assemble into a nanocrystalline liquid crystal phase (NLCP) which was used to template silica. Removal of template yielded a crack-free mesoporous film with layered features that originated from the nematic organization of the ChNCs.

Cellulose is the most abundant natural biopolymer, which is renewable, biodegradable, and non-toxic. It is a polysaccharide composed of monomeric anhydroglucose units connected by β-1,4-glycosidic bonds. Cellulose also exhibits self-assembly properties and an inherent structural hierarchy. It is considered as a green alternative to petroleum-based polymers commonly used as template for the fabrication of mesostructured silica phases. When subjected to acid hydrolysis, cellulose nanocrystals (CNCs), a colloidal suspension, containing the crystalline portion of the polysaccharide, have been reported to assemble into nematic liquid crystalline phase (NLCP), is obtained [21]. This entity, CNC, has been used as a renewable and inexpensive template to cast hybrid mesophase in material synthesis. Dujardin et al. [1] reported a novel synthetic approach to the synthesis of mesoporous silica which involved the sol-gel mineralization of partially ordered suspension of nematic liquid crystalline CNC template. Removal of the CNC template by calcination at 400°C produced a cracked birefringent silica matrix with pattern mesoporosity consisting of co-aligned cylindrical pores of approximately 15 and 9 nm diameter and wall thickness, respectively. Significant extension in pore length far beyond the length of individual CNCs were found to be suggestive of both lateral and lengthwise close packing in the nanocrystal bundles of the nematic template.

The versatility of use of cholesteric phase of cellulose derivatives was demonstrated by Thomas and Antonietti [22] with the use of NLCP of hydroxypropyl cellulose for the generation of a porous silica with high surface area devoid of long-range chiral organization of the template. A freestanding mesoporous silica with high surface area that is a cast of a chiral nematic liquid crystal formed from CNC was reported by Shopsowitz et al. [23]. It was the first material to combine mesoporosity with long-ranged chiral ordering that showed photonic effects. Recently, mesoporous KIE-6 (Korea Institute of Energy-6) was synthesized using crude glycerol waste (CGlW) from biodiesel production. Biodiesel is synthesized through the trans-esterification of triglycerides from biomass such as palm and jatropha; about 1 million tons of CGlW is generated annually. CGlW (consisting of water and methanol, 40%, and glycerol, soap, and fatty acid methyl ester (FAME), 60%) waste constitutes an environmental problem because it is eliminated by incineration, releasing green house gases into the environment, as the cost for its purification is high. Hence, its use in the synthesis of porous material is a solution to its disposal. The glycerol/silica composite with tunable pore size, pore volume, and specific surface area was synthesized by adding glycerol waste and sulfuric acid into the silica sols at room temperature. However, a pre-calcination step at 150°C for 2 h was introduced to prevent the escape of the pore-forming glycerol before the formation of a rigid network by silanol group condensation. Otherwise, glycerol is evaporated before condensation resulting in reduced pore size and volume in the calcined composites. The CGlW template was removed by calcination in air at 550°C for 2 h yielding a 3-D interconnected worm-like mesoporous KIE-6. Properties of mesoporous KIE-6 such as specific surface area, pore size, and volume, and silica wall thickness were tuned by varying the crude glycerol concentration. Variation in glycerol: silica weight ratio in the composite from 0 through 4.9 to 8.0 yielded composites with increased pore size of 4.7 nm through 7.2 and 10 nm, respectively; pore volume from 0.53 to 1.01 cm^3/g and stable Brunauer-Emmett-Teller (BET) surface area between 443 and 550 m^2/g. In addition, life cycle assessment (LCA) of all the indexes included in the synthetic process from biomass cultivation, through biodiesel production to synthesis of mesoporous material, showed that CO_2 emitted during KIE-6 synthesis was much lower than in other existing routes for mesoporous silica. The reduction was attributed to the utilization of renewable CGlW. A major drawback in this synthetic approach, however, is the inclusion of methanol in the CGlW [24].

3.2. Utilization of silica frameworks from biomass residues

Rice, grasses (Germinaecea: bamboo, wheat, maize, and oats), sedges (Cyperaceae), and horsetails (Equisetums) are known to contain high concentration of biogenic silica. Silica is believed to be absorbed from the large abundance of silica in the earth crust released by geological processes as water-soluble silicic acid, $Si(OH)_4$, which through subsequent polymerization and precipitation are deposited as extracellular deposits on cell walls. This process is known as biosilicification. Biosilicification occurs under mild conditions, generating a variety of complex and hierarchical nanostructures silica frameworks which contributes to cell wall rigidity, photosynthesis efficiency, and increased resistance to pathogenic attacks and diseases [25]. Rice husk (RH), which is currently the main natural silica producer, is a waste generated by rice milling activities and this accounts for 20–25 wt% of the whole dried rice grain. According to the United Nations Food and Agriculture Organization [26], the world rice production in 2014–2015 was estimated to be around 740.2 million tons. Therefore, RH makes up about 20 wt%; rice husk waste is a vast silica source. Due to their characteristic woody and abrasive nature, great bulk, slow biodegradability, and low nutritive properties, efforts made to utilize RHs have been very limited. These involve the pre-treatment of the RH to increase its surface area, decrease crystallinity of cellulose, eliminate hemicelluloses and break the lignin seal [27]. RH is generally removed by burning to yield particulate rice husk ash. RHA, containing nearly 90% silica and other metallic oxide as impurities, is commonly used raw materials for obtaining highly pure silica [28]. Trace metallic impurities in RHA silica have been found to be responsible for the clustering of primary silica nanoparticles (Nps). These metallic cations such as K^+ promote the melting of silica Nps and their concentration are responsible for different degrees of melting and different pore structures and pore sizes (**Figure 2**) [29]. Several reports on the extraction of pure biogenic nanosilica from RHA have been published [30–33]. However, before its utilization, a preliminary purification step is required to remove its constituent metallic oxides impurities. After calcination to eliminate the organics, metallic impurities such as oxides of alkali metals are removed by leaching with NaOH solution at temperatures between 100 and 200°C for 1 h. Direct recovery of silica by precipitation from the basic solution is achieved by addition of drops of sulfuric acid to the hot basic solution, followed by thermal treatment at 110°C [34]. Several products of commercial interest such as silicon carbide (SiC) used as semiconductors, pozzolan in cement manufacturing, and recently, sodium silicate in the preparation of silica-based mesoporous materials have been reported [35–42].

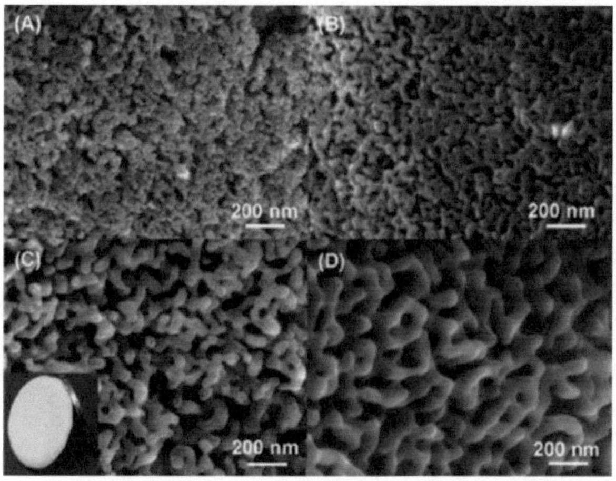

Figure 2. SEM image of nanoporous silica: (A) 0.20 M K^+ pyrolysis at 800°C for 2 h; (B) 0.20 M K^+, pyrolysis at 800°C for 4 h; (C) 0.20 M K^+, pyrolysis at 800°C for 8 h; (D) 0.50 M K^+, pyrolysis at 800°C for 8 h. Effect of K^+ on mesoporous silica from bagasse ash [29].

Siriluk and Yuttapong [42] demonstrated that RHA silica generated from biomass power plant can be used as silica framework for synthesis of MCM-41-type material with crystallinity characteristics and porosity similar to those synthesized from commercial silica. Some authors have reported the synthesis MCM-41-, MCM-48-, and SBA-15-type materials on which functional entities were grafted on the surface silanol groups using sodium silicate solutions extracted from RHA [36, 37, 40]; their potential for CO_2 and chlorinated volatile organic molecules adsorption was found to be similar to those synthesized from conventional silica sources. Recently, Rahman et al. [43] synthesized mesoporous silica with controlled pore structure from bagasse ash with approximately 656 m^2/g surface area and pore diameter 18 nm. Pore structure was influenced by the presence of Na^+, template concentration, and pH.

4. Options to explore the use of both templates and precursors from renewable resources

Several attempts have been made to employ inexpensive, non-toxic, and renewable biomaterials in synthesizing "*green*" porous materials, as the potential of these materials are now being recognized [44]. This is achieved by the extraction of template from biomass and infiltrating it with aqueous inorganic species in solution followed by calcination to remove the template. This approach has been applied in the synthesis of metal carbide replica such as SiC and ZrC [44]. The inherent complexities of organic components present in biological microstructure of biomass after the removal of the inorganic have been employed to create hybrid replicas, thus modifying and enhancing the properties of calcined hybrid material via incorporation of features (such as high surface area, porosity, and even cellular functions) of the biological structure. Ogasawara et al. [45] synthesized ordered porous silica-chitin composite, using well-organized β-chitin organic matrix replica of the chamber-like architecture of cuttlebone. The demineralization of the cuttlebone produced a white organic matrix in which the 3-D chamber-like architecture of the cuttlebone was preserved. Treatment of the cuttlebone template in a sodium silicate solution at pH 11.5 followed by reaction in a series of ethanol-water mixture between pH 10 and 10.5 yielded a porous replica with 3-D interconnected box structures. Witoon et al. [15] reported the synthesis of a bimodal porous silica (BPS) with wormhole-like mesopore with large surface area (783 m^2/g), large total volume, high thermal stability, and tunable pore diameter (3.4–14 nm) from low-cost RHA sodium silicate and deacetylated chitosan as template. The organization and particle size in BPS were found to increase with the increase in pH. Materials with wormhole-like mesopores and macropores were obtained, respectively, at pH 3 and 5 after template removal by calcination. This is due to the increasing solubility and condensation rate at increasing pH and the fact that degree of hydrolysis of the silicate species is at very minimum close to neutral pH, thus forming larger silica particles [46]. A lot of opportunities to synthesis porous materials abound in the environment, particularly those involving the utilization of waste materials. This would without doubt, enhance the value of these wastes as well as proffer a solution to the menace caused by their disposal.

http://dx.doi.org/10.5772/63463

5. Conclusion

The synthesis of mesoporous silica material from commercially available templates and silica frameworks has made the large-scale production of this material unrealistic. This is due to high cost and the negative environmental impacts of the reagents and the harsh synthetic conditions. Recently, attempts have been made to utilize alternative reagents from renewable sources and benign synthetic routes. However, these efforts have largely been directed toward the part utilization of renewable biomass materials as reagents at near neutral pH. Although the materials derived showed characteristics and properties that are similar to those synthesized from commercially available reagents, efforts directed to the full utilization of renewable reagents belittle this fact. Therefore, there is need to focus attention on the full utilization of renewable biomass-generated reagents in the synthesis of MCM. This approach will not only result in the low-cost production of less toxic materials that will be suitable for various applications but also proffer solutions to the environmental and health threats associated with sustainable biomass waste removal.

Acknowledgements

This review was facilitated by the collaboration established between Obafemi Awolowo University (OAU) Ile-Ife and Cape Peninsula University of Technology (CPUT), Cape Town. The authors wish to acknowledge the support of these institutions. TV wishes to acknowledge OAU for providing him a sabbatical position that led to the commencement of this project, while AI acknowledges the support of Joseph Ayo Babalola University for her graduate program.

Author details

Adebola Iyabode Akinjokun[1,3], Tunde Victor Ojumu[2,4*] and
Aderemi Okunola Ogunfowokan[1]

*Address all correspondence to: ojumut@cput.ac.za

1 Department of Chemistry, Obafemi Awolowo University, Ile-Ife, Nigeria

2 Department of Chemical Engineering, Obafemi Awolowo University, Ile-Ife, Nigeria

3 Department of Chemical Sciences, Joseph Ayo Babalola University, Nigeria

4 Department of Chemical Engineering, Cape Peninsula University of Technology, Cape Town, South Africa

References

[1] Dujardin E, Blaseby M, Mann S. Synthesis of mesoporous silica by sol–gel mineralisation of cellulose nanorod nematic suspensions. Journal of Materials Chemistry. 2003;13(4):696–9.

[2] Beck J, Vartuli J, Roth W, Leonowicz M, Kresge C, Schmitt K, et al. A new family of mesoporous molecular sieves prepared with liquid crystal templates. Journal of the American Chemical Society. 1992;114(27):10834–43.

[3] Tan B, Lehmler H-J, Vyas SM, Knutson BL, Rankin SE. Controlling nanopore size and shape by fluorosurfactant templating of silica. Chemistry of Materials. 2005;17(4):916–25.

[4] Yanagisawa T, Shimizu T, Kuroda K, Kato C. The preparation of alkyltriinethylaininonium–kaneinite complexes and their conversion to microporous materials. Bulletin of the Chemical Society of Japan. 1990;63(4):988–92.

[5] Tanev PT, Pinnavaia TJ. A neutral templating route to mesoporous molecular sieves. Science. 1995;267(5199):865–7.

[6] Huo Q, Margolese DI, Ciesla U, Demuth DG, Feng P, Gier TE, et al. Organization of organic molecules with inorganic molecular species into nanocomposite biphase arrays. Chemistry of Materials. 1994;6(8):1176–91.

[7] Israelachvili JN, Mitchell DJ, Ninham BW. Theory of self-assembly of hydrocarbon amphiphiles into micelles and bilayers. Journal of the Chemical Society, Faraday Transactions 2: Molecular and Chemical Physics. 1976;72:1525–68.

[8] Canlas CP, Pinnavaia TJ. Bio-derived oleyl surfactants as porogens for the sustainable synthesis of micelle-templated mesoporous silica. RSC Advances. 2012;2(19):7449–55.

[9] Thomas B, Baccile N, Masse S, Rondel C, Alric I, Valentin R, et al. Mesostructured silica from amino acid-based surfactant formulations and sodium silicate at neutral pH. Journal of Sol-gel Science and Technology. 2011;58(1):170–4.

[10] Rondel C, Alric I, Mouloungui Z, Blanco J-F, Silvestre F. Synthesis and properties of lipoamino acid–fatty acid mixtures: influence of the amphiphilic structure. Journal of Surfactants and Detergents. 2009;12(3):269–75.

[11] Wei Dong X, Yu Xiang Y, Jing C, Zhaolun W, Xiang Nong L. Preparation of mesoporous silica using amphoteric surfactant potassium and sodium N-dodecyl glycine template. Journal of the American Ceramic Society. 2008;91(5):1517–21.

[12] Štangar UL, Hüsing N. Alkyl-glycoside surfactants in the synthesis of mesoporous silica films. Silicon Chemistry. 2003;2(3–4):157–65.

[13] El Kadib A, Bousmina M. Chitosan bio-based organic–inorganic hybrid aerogel microspheres. Chemistry-A European Journal. 2012;18(27):8264–77.

[14] Pedroni V, Schulz P, de Ferreira MG, Morini M. A chitosan-templated monolithic siliceous mesoporous-macroporous material. Colloid and Polymer Science. 2000;278(10):964–71.

[15] Witoon T, Chareonpanich M, Limtrakul J. Synthesis of bimodal porous silica from rice husk ash via sol–gel process using chitosan as template. Materials Letters. 2008;62(10): 1476–9.

[16] Alonso B, Belamie E. Chitin–silica nanocomposites by self assembly. Angewandte Chemie International Edition. 2010;49(44):8201–4.

[17] Monnier A, Schüth F, Huo Q, Kumar D, Margolese D, Maxwell R, et al. Cooperative formation of inorganic-organic interfaces in the synthesis of silicate mesostructures. Science. 1993;261(5126):1299–303.

[18] Pouget E, Dujardin E, Cavalier A, Moreac A, Valéry C, Marchi-Artzner V, et al. Hierarchical architectures by synergy between dynamical template self-assembly and biomineralization. Nature Materials. 2007;6(6):434–9.

[19] Nguyen TD, Shopsowitz KE, MacLachlan MJ. Mesoporous silica and organosilica films templated by nanocrystalline chitin. Chemistry–A European Journal. 2013;19(45): 15148–54.

[20] Revol J-F, Godbout L, Dong X-M, Gray DG, Chanzy H, Maret G. Chiral nematic suspensions of cellulose crystallites; phase separation and magnetic field orientation. Liquid Crystals. 1994;16(1):127–34.

[21] Habibi Y, Lucia LA, Rojas OJ. Cellulose nanocrystals: chemistry, self-assembly, and applications. Chemical Reviews. 2010;110(6):3479–500.

[22] Thomas A, Antonietti M. Silica nanocasting of simple cellulose derivatives: towards chiral pore systems with long range order and chiral optical coatings. Advanced Functional Materials. 2003;13(10):763–6.

[23] Shopsowitz KE, Qi H, Hamad WY, MacLachlan MJ. Free-standing mesoporous silica films with tunable chiral nematic structures. Nature. 2010;468(7322):422–5.

[24] Lee D-W, Jin M-H, Park JC, Lee C-B, Oh D, Lee S-W, et al. Waste-glycerol-directed synthesis of mesoporous silica and carbon with superior performance in room-temperature hydrogen production from formic acid. Scientific reports. 2015;1–5.

[25] Wang W, Martin JC, Zhang N, Ma C, Han A, Sun L. Harvesting silica nanoparticles from rice husks. Journal of Nanoparticle Research. 2011;13(12):6981–90.

[26] Rice Market Monitor [http://www.fao.org/fileadmin/templates/est/COMM_MARKETS_MONITORING/Rice/Images/RMM/RMM_DEC15_H.pdf]. 2015.

[27] Lim JS, Manan ZA, Alwi SRW, Hashim H. A review on utilisation of biomass from rice industry as a source of renewable energy. Renewable and Sustainable Energy Reviews. 2012;16(5):3084–94.

[28] Lu P, Hsieh Y-L. Highly pure amorphous silica nano-disks from rice straw. Powder Technology. 2012;225:149–55.

[29] Wang W, Martin JC, Fan X, Han A, Luo Z, Sun L. Silica nanoparticles and frameworks from rice husk biomass. ACS Applied Materials & Interfaces. 2012;4(2):977–81.

[30] Adam F, Appaturi JN, Iqbal A. The utilization of rice husk silica as a catalyst: review and recent progress. Catalysis Today. 2012;190(1):2–14.

[31] Chuah T, Jumasiah A, Azni I, Katayon S, Choong ST. Rice husk as a potentially low-cost biosorbent for heavy metal and dye removal: an overview. Desalination. 2005;175(3):305–16.

[32] Yalcin N, Sevinc V. Studies on silica obtained from rice husk. Ceramics International. 2001;27(2):219–24.

[33] Zerbino R, Giaccio G, Isaia G. Concrete incorporating rice-husk ash without processing. Construction and Building Materials. 2011;25(1):371–8.

[34] Kalapathy U, Proctor A, Shultz J. A simple method for production of pure silica from rice hull ash. Bioresource Technology. 2000;73(3):257–62.

[35] An D, Guo Y, Zhu Y, Wang Z. A green route to preparation of silica powders with rice husk ash and waste gas. Chemical Engineering Journal. 2010;162(2):509–14.

[36] Bhagiyalakshmi M, Lee JY, Jang HT. Synthesis of mesoporous magnesium oxide: its application to CO_2 chemisorption. International Journal of Greenhouse Gas Control. 2010;4(1):51–6.

[37] Bhagiyalakshmi M, Yun LJ, Anuradha R, Jang HT. Synthesis of chloropropylamine grafted mesoporous MCM-41, MCM-48 and SBA-15 from rice husk ash: their application to CO_2 chemisorption. Journal of Porous Materials. 2010;17(4):475–84.

[38] Boissière C, Larbot A, Prouzet E. Synthesis of mesoporous MSU-X materials using inexpensive silica sources. Chemistry of Materials. 2000;12(7):1937–40.

[39] Chumee J, Grisdanurak N, Neramittagapong S, Wittayakun J. Characterization of AlMCM-41 synthesized with rice husk silica and utilization as supports for platinumiron catalysts. Brazilian Journal of Chemical Engineering. 2009;26(2):367–73.

[40] Grisdanurak N, Chiarakorn S, Wittayakun J. Utilization of mesoporous molecular sieves synthesized from natural source rice husk silica to chlorinated volatile organic compounds (CVOCs) adsorption. Korean Journal of Chemical Engineering. 2003;20(5): 950–5.

[41] Jang HT, Park Y, Ko YS, Lee JY, Margandan B. Highly siliceous MCM-48 from rice husk ash for CO_2 adsorption. International Journal of Greenhouse Gas Control. 2009;3(5): 545–9.

[42] Siriluk C, Yuttapong S, editors. Structure of mesoporous MCM -41 prepared from rice husk ash. 8th Asian symposium on visualization; 2005; Chiangmai, Thailand.

[43] Rahman NA, Widhiana I, Juliastuti SR, Setyawan H. Synthesis of mesoporous silica with controlled pore structure from bagasse ash as a silica source. Colloids and Surfaces A: Physicochemical and Engineering Aspects. 2015;476:1–7.

[44] Schnepp Z. Biopolymers as a flexible resource for nanochemistry. Angewandte Chemie International Edition. 2013;52(4):1096–108.

[45] Ogasawara W, Shenton W, Davis SA, Mann S. Template mineralization of ordered macroporous chitin-silica composites using a cuttlebone-derived organic matrix. Chemistry of Materials. 2000;12(10):2835–7.

[46] Brinker CJ, Scherer GW. Sol-gel scienceAcademic Press, San Diego, CA, 1990.

Chapter 6

Overview of Phosphorus Effect in Molybdenum-Based Hydrotreating Catalysts Supported on Ordered Mesoporous Siliceous Materials

Rafael Huirache-Acuña, Eric M. Rivera-Muñoz,
Trino A. Zepeda, Rufino Nava and Barbara Pawelec

Additional information is available at the end of the chapter

http://dx.doi.org/10.5772/64181

Abstract

This chapter presents an overview of the literature on the effect of phosphorus modification of ordered mesoporous silica (OMS) such as MCM-41, HMS, SBA-15, and SBA-16 to be used as supports for hydrotreating catalysts based on transition metal sulfides (TMS). The influence of the support modification with variable quantities of phosphorus on the performance for hydrodesulfurization (HDS) and hydrodenitrogenation (HDN) reactions is outlined and discussed considering the changes in the structural and textural properties (S_{BET}), acidity, reducibility, etc., of the substrate brought about by phosphorus incorporation.

Keywords: phosphorus, hydrotreating, catalysts, mesoporous, siliceous materials

1. Introduction

Heterogeneous catalysts based on transition metal sulfides (TMS) used in the refinery for the hydrotreatment of middle distillates are usually supported on an alumina substrate [1]. This is because alumina is of low cost and shows remarkable textural and mechanical properties [2]. However, as the alumina-supported hydrotreating catalysts display only moderate acidity, it is a common practice to enhance it by surface grafting with H_3PO_4 [3]. Besides the enhancement of the support's acidity, the phosphorus might act as a promoter [4]. Unfortunately, alumina-supported catalysts exhibit undesirable strong metal-support interaction (SMSI) leading to the formation of $AlPO_4$ phase [3]. To overcome this problem, a common practice is phosphatation of the alumina surface which reduces the strong interactions of the molybdenum fraction with the alumina surface [5]. The addition of H_3PO_4 to the impregnation solution improves stability and increases metal solubility which enables one to prepare binary Co(Ni)-Mo(W) catalysts in a single impregnation step. Phosphate has been reported to improve the HDS and especially HDN activities of the Mo-based catalysts [3]. The explanations on the beneficial effects of the presence of phosphate validate that phosphate acts as a second promoter [6], decreases the metal-support interactions, improves the solubility of molybdate by the formation of phosphomolybdate complexes which led to easier catalyst preparation [6], inhibits the formation of the inactive $CoAl_2O_4$ or $NiAl_2O_4$ species [5], and favors the formation of the Type-II Co(Ni)-Mo-S structures [7].

The effect of phosphorus strongly depends on P loading, catalyst preparation method, and support chemistry and its morphology [8]. All these factors might explain the controversial catalytic behavior of P-modified Mo-based catalysts in hydrodesulfurization (HDS), hydrodenitrogenation (HDN), and hydrogenation (HYD) reactions. A comprehensive review of the physicochemical processes that occur on the surface of the alumina-supported catalysts modified with phosphorus was carried out by Iwamoto and Grimblot [3]. Recently, the state-of-the art overview of the literature on the effects of phosphorus promotion and poisoning in aluminosilicates (zeolites) was conducted by van der Bij and Weckhuysen [9]. However, to the best of our knowledge, the overview on phosphorus promotion and poisoning in bare silica substrates has not been reported yet. This is probably because compared to alumina-based catalysts, the effect of phosphorus modification of ordered mesoporous silica (OMS) was scarcely investigated [10–17]. Ordered mesoporous silica is a substrate worthy of note as it supports active phases for hydrotreating catalysts. This is because of their interesting textural properties such as a high surface area (above 1000 $m^2 g^{-1}$), pores in the range of mesopores (2–50 nm), and complementary textural porosity. As the pore wall surface of OMS is carpeted with a high concentration of silanol groups, it can be easy functionalized with different functional groups. As it occurs with alumina-supported catalysts, it was found that the effect of phosphorus modification of the OMS strongly depends on not only P loading but also the method of phosphorus incorporation. Notwithstanding, it is emphasized that the effect of support modification with phosphorus on the catalytic performance is not clearly understood and the activity results are often contradictory. In some cases, the promoting effect of phosphorus on the HDS activity was observed [10–12, 15, 17], while the inhibition effect of phosphorus was found in other studies [16]. OMS exhibits ordered arrangements of channels and/or cavities of different geometries built up from SiO_2 units. Their pore size can be controlled and modified, in a reasonable range, using in situ and ex situ synthetic strategies. The effect of P was investigated [11] for the ordered mesoporous silicate, called M41S, which was first synthesized at the Mobil Corporation in 1992 [18]. This material shows a large specific area, a hexagonal array, and uniform pore channels, yet its practical application is limited due to its poor hydrothermal stability. Subsequent to MCM-41, the synthesis of other hexagonal mesoporous materials (HMS) was reported [19, 20]. In comparison to MCM-41, the textural characteristics of the HMS material have certain advantages due, in part, to its larger textural mesoporosity and wormhole mesostructure (*vide infra* **Figure 1**) which offers better transport for reactants and products [15]. The HMS framework structures, particularly those with a sponge-like particle texture, were claimed to be promising materials for supporting heterogeneous catalysts [12, 21, 22].

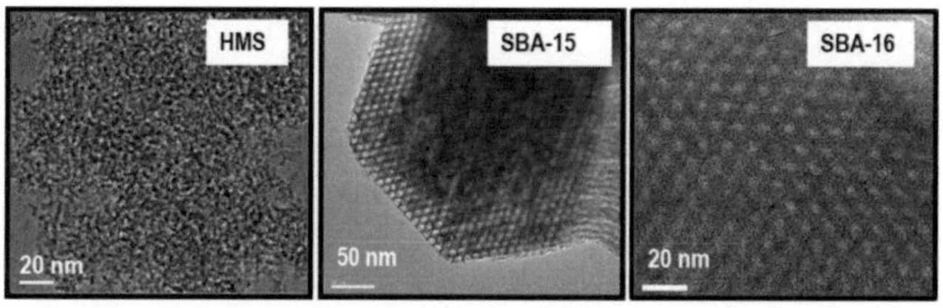

Figure 1. HRTEM images of the HMS substrate showing wormhole structure and SBA-15 and SBA-16 substrates displaying highly ordered hexagonal pores in a 2D array.

The more remarkable advance in the synthesis of mesoporous systems was reported by Zhao et al. [23], who synthesized hydrothermally stable mesoporous SBA-15 silica molecular sieves using triblock copolymer surfactants as template. The calcined SBA-15 demonstrated to be more hydrothermally stable than its HMS counterpart with wormhole structure because of its thicker walls with a network of micropores within the walls. The mesoporous SBA-15 has hexagonal pores in a two-dimensional (2D) array with long 1D channels corresponding to the *P6mm* space group symmetry (*vide infra* **Figure 1**). Thus, similar to the case of MCM-41, the parallel channel system of SBA-15 might act as a microreactor, wherein the reactant and the intermediary products will be in prolonged contact with the active phase [2]. In addition to OMS with cylindrical mesopores, the mesoporous silica with a cage-like structure such as SBA-16 has been used for supporting hydrotreating catalysts [24–26]. The siliceous SBA-16 exhibits an *Im3m* space group symmetry (*vide infra* **Figure 1**) with three-dimensional (3D) structure formed by spherical cavities arranged in a body-centered cubic array, which are connected through eight apertures to the nearest neighbors along the (1 1 1) directions [27]. As a consequence, the interconnected spherical mesopores are easily accessible for guest molecules, which facilitate the transport of reactants and products without pore blockage [23, 27].

In this chapter, the effect of phosphate modification of mesoporous silica substrates such as MCM-41, HMS, SBA-15, and SBA-16 on the catalytic response of transition metal sulfide catalysts in hydrotreating reaction has been examined. As most of our conclusions on the effect of OMS modification with phosphorus agree with those discussed recently for aluminosilicates, readers are suggested to study the review by van der Bij and Weckhuysen [9].

2. Synthesis of P-containing OMS

Similar to alumina and alumina-silicate-based materials [3, 9], the method of phosphorus introduction was found to play an important role in the distribution of phosphorus in the mesoporous siliceous systems [28, 29]. Two strategies have been undertaken to modify OMS with phosphorus: (i) direct sol-gel, and (ii) post-synthesis methods. For the pure siliceous substrates, phosphorus incorporation by post-synthesis grafting was more frequently employed than sol-gel synthesis because of its low cost and simplicity.

Direct sol-gel P introduction into OMS was more often employed in medical science applications than in petrochemistry. For example, Vallet-Regi et al. [30] synthesized a P-containing MCM-41 mesoporous material via one-step sol-gel preparation method and then applied as a bioactive material. The characterization of this material by different techniques demonstrated that only a small amount (under 1% P) was incorporated into the structure of MCM-41. The P-containing MCM-41 sample prepared via sol-gel exhibited the characteristic ordered hexagonal array of mesopores of this kind of materials.

Direct sol-gel synthesis was also employed by Pitchumani et al. [28] for the preparation of the SBA-15 substrates with distinct morphologies (well-ordered mesoporous films, cakes, fibers, and bundle-like structures). In general, the synthesis of the SBA-15 substrate involves the formation of organic-inorganic composites by a self-assembly process in which the organic phase serves as a template for the inorganic phase [23]. In this process, tetra-ethyl orthosilicate (TEOS) is used as silica source, and a nonionic P123 block copolymer as surfactant (EO_{20}-PO_{70}-EO_{20}, where EO is ethylene oxide and PO is propylene oxide) and a self-assembly of both phases occurs in the presence of strong acids such as HCl.

With the aim to elucidate the reaction between the phosphate ions and silica species, Tagaya et al. [29] investigated the influence of the P incorporation methodology (coating vs. doping) on the nanostructure of phosphorus-containing mesoporous silica films. Phosphoric acid was added before and after the mesopore formation (doping and coating methods, respectively). In the coating method, the calcined mesoporous silica film was spin-coated with an aqueous solution of H_3PO_4 at four different rotation speeds and then calcined at 450°C for 6 h. It was observed by XRD that the films prepared by the coating method exhibited high-ordered pore channels with a hexagonal structure parallel to the substrate surface, while those prepared by doping method exhibited larger pores with worm-like structures. This could be explained considering that the coating method leads to the interaction of H_3PO_4 with the silica species. In such cases, the P-O-Si bonds can be formed during hydrogenolysis, while the hydroxyl groups of phosphoric acid can be eliminated by calcination. Besides, the hydroxyl groups on phosphorus species can be restored by exposing the calcined sample to an air mixture [28]. In the doping method, an aqueous solution of phosphoric acid is added instead of HCl during the mesoporous silica film preparation (one-pot synthesis). In such cases, a larger amount of H_3O^+ and phosphate ions could be adsorbed on the hydrophilic head groups of the cationic surfactant of cetyltrimethylammonium chloride (CTAC) leading to the formation of Si-O-P-OH groups [29].

Using the post-synthesis method, phosphorus could be grafted onto the surface of silica materials by treating with phosphoric acid for a very short time, followed by high-temperature calcination [31]. This method was employed by many researchers. For example, it was used for phosphorus incorporation onto the surface of SBA-16 [25], HMS [15], Al-HMS [22], MCM-51, and SBA-15 substrates [9]. Using the post-synthesis grafting method for phosphorus introduction, the bare mesoporous silica substrate is synthesized first and after its calcination, the pure siliceous material is made to be in contact with an aqueous solution of

H_3PO_4 having an appropriate concentration of phosphoric acid. After drying, the obtained materials are calcined. For example, for the preparation of P-containing HMS, the S°I° assembly of HMS molecular sieve silica with wormhole framework structures can be accomplished using dodecylamine (DDA) as the structure-directing surfactant (S°) and tetra-ethyl orthosilicate (TEOS) as the inorganic precursor (I°) [32]. The procedure similar to that reported by Zhang et al. [23] was employed by Nava et al. [15] for the preparation of pure siliceous HMS material. After solid calcination at 813 K for 6 h, the P-modified HMS was prepared by making the parent HMS come in contact with an aqueous solution of H_3PO_4 (incipient wetness impregnation). Then, the solid was dried at 383 K overnight and then calcined at 773 K for 3 h under static conditions. In the case of the P-containing HMS-Al, the parent Al-HMS substrate (Si/Al molar ratio of 40) was prepared by Zepeda et al. [22] following the procedure described by Gontier and Tuel [33].

Finally, another variation of the post-synthesis method for P-incorporation is the simultaneous impregnation of the bare support with metal salt precursors and H_3PO_4. This method, which is commonly used for the manufacture of HDS catalysts at commercial scale, was employed by Hernández et al. [34] for the preparation of Co-Mo-P/Al-MCM-41 catalysts.

3. Influence of phosphorus on morphology of OMS materials

In general, the incorporation of phosphate on the OMS substrates by post-synthesis grafting method does not change their structure, as it was confirmed by comparison of small-angle XRD patterns of the P-loaded and unloaded HMS [15 ,21], SBA-15, and SBA-16 [24]. This is probably because the separate P_2O_5 phase could be formed during calcination of the acid-functionalized solids having $-PO_3H$ or $-PO_2H$ groups formed through interaction of the P-OH bonds of H_3PO_4 and the surface silanols [31]. These phosphate species remain deposited as a separate phase on the surface of OMS substrates, as it was confirmed for P/HMS systems by Nava et al. [15]. This is not the case of more reactive alumina for which phosphate reacts to a large extent on the surface layers of the alumina forming a well-dispersed $AlPO_4$ phase [6, 35]. In the case of P incorporation on the surface of calcined Ti-HMS, the textural and surface properties of HMS were modified by Ti cations incorporated during one-pot synthesis. The physical and chemical characterization of such calcined substrates indicated that the presence of phosphorus pentoxide species on the support surface does not change the mesoporous character of the HMS-Ti substrate, but modifies its surface properties [21]. The P/HMS-Ti substrates, which were prepared by wet impregnation of HMS-Ti substrate with H_3PO_4 followed by calcination at 500°C, exhibited almost a linear decrease in the specific surface area (S_{BET}) with increasing P loading. This is because of blocking of some support pores by P_2O_5 species located almost exclusively on the external support surface, as demonstrated by calculation of the normalized BET surface area [21]. A higher concentration of phosphorus species on the zeolite surface compared to phosphorus in the bulk was also reported for the P-modified zeolites, wherein the introduction of phosphoric acid into a zeolite was made by wet impregnation [36–38]. This is because dehydration occurring during heat treatment leads to the formation of large condensed polyphosphate and P_2O_5 species [35]. This is in good agreement with the results of the P-modified zeolites [9]. The majority of these phosphorus species exhibited no interaction with mesoporous silica, which are considered as separate phases [16, 17, 21]. Using the small-angle X-ray powder diffraction technique, Zepeda et al. [21] confirmed that all P(*x*)/HMS-Ti substrates modified with small amounts of P (0.2–1.2 wt. % of P_2O_5) exhibited single small-angle reflection corresponding to the (100) plane, which is characteristic of the HMS material. Although higher-order Bragg reflections were not observed in the patterns of S°I° HMS materials, the single d_{100} peak indicates that all materials display a short-range hexagonal symmetry with a uniform pore diameter. A similar XRD pattern was reported for the HMS-Ti system [33]. For the P-free HMS-Ti substrate, the small-angle reflection is located at $2\theta = 1.30°$. After P loading onto the surface of HMS-Ti, the d_{100} peak is shifted to higher angles (from 1.30 to 1.38°). The calculation of the wall thickness indicates a slight decrease in this parameter after addition of P to HMS-Ti (from 4.27 to 3.92 nm). This may indicate that impregnation of HMS-Ti substrate with H_3PO_4 led to a lattice reduction and a decrease in pore diameter. Indeed, the unit cell parameter (a_0) calculated using the equation $a_0 = 2d_{100}/\sqrt{3}$ indicated that support modification with phosphorus produced a slight decrease in the unit cell parameter (from 7.67 to 7.12 nm) [21]. Similar to the phosphorus effects, an increase in the pore diameter might occur if the catalyst is prepared by the simultaneous impregnation method as compared with that prepared by the successive impregnation [13]. This is deduced from **Figure 2** which shows the XRD patterns of the HMS-based catalysts prepared by simultaneous and successive impregnation. As seen in this figure, the oxide precursors prepared by simultaneous and successive impregnation methods exhibit their d_{100} peaks around 1.14 and 1.34° in 2θ, thus suggesting an increase in the dimensions of the scattering domain, and in consequence, an increase of the pore diameter for the samples prepared by simultaneous impregnation [13].

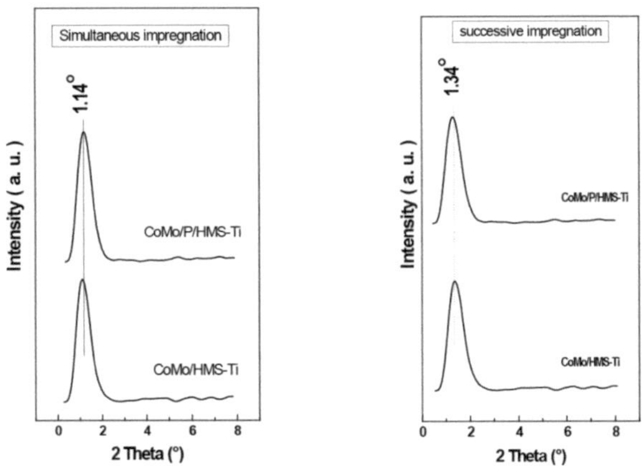

Figure 2. Low-angle X-ray diffraction patterns of the oxide precursors of CoMo/HMS-Ti and CoMo/P/HMS-Ti catalysts prepared by successive and simultaneous impregnation methods [13] (reproduced with permission from Elsevier).

Similar to the P/HMS-Ti-based catalysts [13, 21], P incorporation on the surface of SBA-15 by post-synthesis grafting did not lead to significant changes in the small-angle XRD patterns of the P/SBA-15 substrates [24]. Moreover, similar to the HMS-based substrates [21], the diffraction peaks of the P-loaded SBA-15 samples were shifted to larger 2θ values with respect to bare SBA-15 samples, thereby indicating that the P-loaded sample shows smaller unit cell parameter than the unloaded one [24]. This means that when phosphoric acid was added to SBA-15, the structural integrity of the mesoporous silica decreased upon increasing the amount of grafted acid. For very low P loading, the uniformly arranged hexagonal mesoporous framework was preserved. Indeed, Shon et al. [39] observed that the hexagonal mesoporous framework of SBA-15 substrate was maintained even after loading of excess of H_3PO_4 onto the surface of this material (Si/P ≈ 10).

4. Changes in textural properties of OMS induced by phosphorus

In general, textural characterization of the P-containing mesoporous silica from the N_2 adsorption-desorption isotherms indicated that P loading by post-synthesis grafting has a detrimental effect on their textural properties, as demonstrated for the P/HMS and P/HMS-Al prepared by grafting the respective parent substrate with H_3PO_4 [17, 22], The observed decrease in the specific BET surface area and pore volume was explained in terms of a partial blockage of the pores by the P species. Similarly, P loading onto siliceous MCM-41 by post-synthesis grafting method caused a deterioration of the textural characteristics and some loss in the periodicity of the MCM-41 pore structure [11]. In order to understand the mechanism of such alteration, the stability of various structured mesoporous materials (SBA-1, SBA-3, SBA-15, MCM-41, and MCM-48) under acidic conditions of H_3PO_4 was investigated [40]. These authors reported that the micropores were first damaged, and then the mesopores were partially collapsed or partially blocked. The resistance of the material against acidic media was associated with a critical wall thickness-pore diameter threshold. Contrary to the grafting method [22], the P incorporation by direct sol-gel synthesis method led to an increase in the specific surface area and pore volume of mesoporous silica substrates for the mesostructurated P-doped silica monoliths. As the P-loaded monolith exhibited a smaller average pore size and unit cell parameter than its unloaded counterpart, the observed increase of specific surface area and pore volume was explained due to the formation of disordered and interconnected pores inside each microdomain [41].

Hexagonal mesoporous HMS (Si/Ti molar ratio of 40) modified with both Ti and P were used as supports for Co(Ni)Mo sulfide phases by Zepeda et al. [21]. The Ti-loaded HMS (Si/Ti molar ratio of 40) was prepared by direct synthesis, wherein the titanium precursor was added to the synthesis gel. After calcination, the Ti^{4+} ions were incorporated into the framework of the HMS substrate. The P-containing supports were prepared by typical impregnation of the calcined HMS-Ti substrate with an aqueous solution of H_3PO_4. Investigation of textural properties of the bare supports demonstrated that Ti insertion into the framework of HMS did not change the mesoporous character but modified the surface properties even though the low-angle XRD patterns demonstrated almost no changes in pore ordering upon P loading and/or preparation procedure [13]. Contrary to P/Ti-HMS [21], the N_2 adsorption-desorption and small-angle XRD measurements of the HMS modified with both Al and P confirmed a decrease of specific surface area and partial destruction of the support mesoporous structure after modification of calcined Al-HMS with phosphorus [12, 22].

5. Metal oxides formed on the support surface modified with P

It is well known that the dispersion and structure of Mo species is a consequence of the modification of the substrate's isoelectric point and a change in the interaction of the adsorption of molybdate species during the impregnation procedure. Thus, depending on the acid concentration, the support impregnation with H_3PO_4 might change the isoelectric point of the support, thereby influencing the interaction of the molybdate species with support. In addition, the ordered framework of the OMS may control the metal oxide particle size limiting the growth of the clusters introduced into the confined space of the channels. Thus, the presence of phosphate species on the support surface might influence the metal oxide location of the phases formed. In this context, the XRD, DRS, and TPR characterizations of the Ni/Mo/P/MCM-41 oxide precursors by Herrera et al. [11] showed that the characteristics (dispersion, coordination state, and temperature of reduction) of oxide Mo and Ni species changed with phosphorus incorporation in the MCM-41.

The influence of phosphorus species on the formation of metal oxide phases on the support surface was investigated by Nava et al. [15, 17]. The X-ray diffraction patterns of the calcined catalysts confirmed the formation of the crystalline MoO_3 phase (JCPDS 5-508) and β-$CoMoO_4$ phase (JCPDS 21-0868). Considering the crystal size of the β-$CoMoO_4$ phase, these authors concluded that the catalysts with low P loading exhibited higher dispersion of Co and Mo phases than their counterparts having high P loadings. The catalyst modified with the largest P loading displayed the largest crystal size of the β-$CoMoO_4$ phase (14.2 nm). This was explained in terms of the decrease of the metal-support interaction induced by a large amount of phosphate species located on the support surface [17]. The wide-angle XRD diffraction technique was employed by Pawelec et al. [13], for characterization of the oxide precursors of CoMo/P/HMS-Ti catalysts prepared by simultaneous and successive impregnation methods. All the samples prepared by simultaneous impregnation did not show reflections belonging to cobalt and molybdenum oxide, suggesting that the phases formed are amorphous or not fully crystalline, or their crystal sizes are below the detection limit of the XRD technique. By contrast, the XRD diffraction patterns of the catalysts prepared by simultaneous impregnation exhibited a small reflection peak at 22.7° 2θ and an intense reflection peak at 25.7° 2θ, overlapping with a broad diffraction of the substrate, which are due to crystalline MoO_3 phase (JCPDS card 1-076-1003). The intensity of both reflections increased suggesting a higher preference of molybdenum ions for P-OH groups of P(x)/HMS-Ti support than for cobalt [42].

6. Effect of phosphorus on the formation of active sites

As mentioned in Section 1, the effect of phosphorus on the HDS activity of catalysts depends on the P loading. The effect of support modification by phosphorus on the morphology of MoS_2

and "Co(Ni)-Mo(W)-S" type structures formed after precursor sulfidation was largely investigated. As the most active catalysts in HDS reaction are mixed "Co(Ni)-Mo(W)-S" structures, the main question arises: ***May the separate phosphate species located on the support surface influence the formation of "Co(Ni)-Mo-S" active phases?*** The most intuitive explanation for the phosphorus inhibition of HDS reaction over supported Co(Ni)Mo(W) catalysts is that the phosphorus species may act as a physical barrier inhibiting the formation of this phase. However, surface analysis by XPS of fresh sulfided CoMo/HMS and CoMo/P/HMS catalysts by Nava et al. [16] strongly suggested that the catalyst preparation method should be more important for the formation of the mixed "Co-Mo-S" phase than the presence of phosphate species on the support surface. Both CoMo/HMS and CoMo/P/HMS sulfide catalysts showed the S 2p peak characteristic of S^{2-} ions (binding energy (BE) of 161.2 ± 0.1 eV). For the CoMo/P/HMS sulfide catalyst, the P 2p peak (BE at 134.0 eV) was indicative of the presence of phosphate species on the support surface. Moreover, the Mo $3d_{5/2}$ core-level spectra appeared at binding energy characteristic of MoS_2 phase (BE at 228.7 eV) and the Co $2p_{3/2}$ core level spectra showed two contributions, one due to the formation of Co_9S_8 phase (BE at 778.2–778.8 eV) and another due to non-sulfided Co^{2+} species (BE at 780.2–780.8 eV). Taking into account these XPS data of both P-containing and P-free catalysts, the formation of "CoMoS" phase is precluded. Thus, it is impossible to conclude that phosphate species on the support surface does not inhibit the formation of this phase. Although the formation of "Co-Mo-S" phase did not occur, the sulfide CoMo/P/HMS catalyst showed surface exposure and sulfidation degree larger for both cobalt and molybdenum than on its P-free counterpart. Thus, the presence of the phosphate species on the support surface enhanced the surface exposure of MoS_2 particles, in close agreement with the HRTEM characterization data [16].

The genesis of the active phases formed during sulfidation of Ni/P/SiO_2 catalysts with H_2/H_2S gas mixture was studied by Koranyi [10]. Under the reaction conditions, the P-containing catalysts exhibited a higher HDS activity than the P-free sulfided Ni/SiO_2. The increase in activity observed for P-containing catalysts was linked with the formation of both nickel sulfide and nickel phosphide phases during catalyst activation during sulfidation. The larger intrinsic activity of the nickel phosphide with respect to nickel sulfide is linked with its different structure. Contrary to the metal sulfides, the structure of the metal phosphides is based on trigonal prisms having large phosphorus atoms (atomic radius of phosphorous is 0.109 nm) located in their centers. Importantly, contrary to the transition metal sulfides, transition metal phosphides do not exhibit a layered structure, which offers potentially better exposure of surface atoms to reactants [43]. The work by Koranyi [10] demonstrated that the simple impregnation procedure was suitable to bring at least part of the nickel in close contact with phosphorus. Contrary to Ni/P/SiO_2, the activity of P-free Ni/SiO_2 catalyst was associated with the nickel sulfide surface area, as confirmed by dynamic chemisorption measurements (DOC). In addition, Koranyi [10] observed that the catalyst presulfided at atmospheric pressure was more active than its counterpart sulfided at an elevated pressure. This result suggests that the type of active phases formed (nickel sulfides or nickel phosphides) depends on the H_2/H_2S gas pressure employed for catalyst sulfidation. The ability of phosphorus to modify the morphology of the Mo(W)S_2 phase has been observed for sulfided NiMoW/SBA-16 catalysts modified with phosphorus by Guzmán et al. [25]. In the light of HRTEM results, the low HDS activity of P-free NiMoW/SBA-16 sample was explained on account of the development of poorly active "onion-type" $Mo(W)_2$ structures (see **Figure 3**). This was also confirmed by HRTEM investigation by Herrera et al. [11] on the morphology of MoS_2 phase formed on the surface of sulfided NiMo/P/MCM-41 catalyst loaded with a low amount of phosphorus (2 wt.% P_2O_5). HRTEM images indicated the formation of small MoS_2 crystallites (0–6 nm) with two or three layers. For these catalysts, an increase of P loading from 2.0 to 5 wt.% led to an increase in the average length and layer stacking of MoS_2 crystallites.

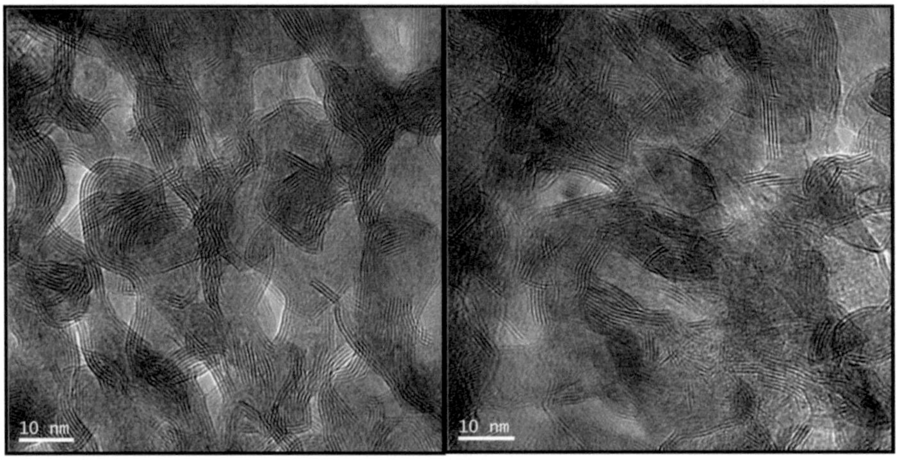

Figure 3. HRTEM images showing the enhancement of the active phase dispersion on the fresh sulfided NiMoW/P/SBA-16 catalyst modified with P with respect to P-free NiMo/SBA-16 (left) (adapted from Guzman et al., [25], with permission of Elsevier).

7. Effect of phosphorus addition on acid sites

The naked mesoporous silicas exhibit electronically neutral framework and lack of Brønsted acidity. As bifunctional catalysts having both metal and acid functions are required for hydrotreating reactions, the formation of Brønsted acid sites on the surface of mesoporous silicas is very important for their potential application as supports of the hydrotreating catalysts. For instance, the bifunctional catalysts are required for the reaction of dehydration of isopropanol [31], hydrodesulfurization of 4,6-Dimethyldibenzothiophene (4,6-DMDBT) [17, 22], and hydrodesoxygenation of anisole [26]. To circumvent this limitation, many approaches have been undertaken, among which the substitution of Si^{4+} by different cations, functionalization with different groups, carrying the preparation method or changing the active phase component remain prominent [44]. In order to control both acid strength and acid site density, functionalization of the silica surfaces with various kinds of organosilane compounds was the most commonly employed methodology [39].

Pitchumani et al. [28] employed a weak acid (H_3PO_4) instead of strong HCl acid during the direct synthesis of P-containing SBA-15 substrates. Assuming the interaction of H_3PO_4 on the silica surface which leads to phosphorus incorporation into the silica framework, Brønsted acid sites are expected to occur. Unfortunately, no experimental evidence of the formation of the Si-O-P bonds was reported. Similarly, Shon et al. [39], claimed the formation of Brønsted acid sites on the SBA-15 catalysts during interaction of the H_3PO_4 at the silica surface. SBA-15-based catalysts were functionalized with -PO_3H_2 groups by post-grafting with H_3PO_4, followed by drying in a vacuum oven at 60°C for 12 h and then by calcination at 500°C for 4 h under static air conditions. The latter process led to the decomposition of phosphorus precursors into (poly)phosphates and phosphorus pentoxide.

Unfortunately, in the studies by Pitchumani et al. [28] and Shon et al. [39], there is no experimental evidence on the formation of those bonds. The same situation occurs in the case of the microporous aluminosilicates [9]. On the contrary, there is experimental evidence on the formation of Lewis acid sites as it was demonstrated by the FTIR spectra of adsorbed pyridine for freshly sulfided P-loaded CoMo/HMS catalysts [15]. Interestingly, the temperature-programmed desorption patterns of desorbed ammonia performed on the oxide precursors of these catalysts demonstrated that the phosphate phase remained on the surface of HMS. These phosphate moieties are responsible for the formation of weak acid sites whose proportion increases with increasing P loading on the substrate [15]. This is in good agreement with the work by Wang et al. [45] who observed that relatively more strong and medium strength acid sites are affected by phosphatation than weak acid sites [17]. Calcined CoMo/HMS catalysts loaded with variable amounts of phosphorus (0.5, 1.0, 1.5, and 2.0 wt.% of P_2O_5) were characterized by ^{31}P NMR spectroscopy by Nava et al. [17]. The ^{31}P NMR spectra indicated that the catalyst with the lowest P loading exhibited isolated phosphate units, whereas the one with P_2O_5 loading of 1.5 wt.% shows the phosphorus atoms connected to silica forming the O=P(OP/OSi) linkages. The catalysts with P_2O_5 loading of 1.0 and 2.0 wt.% resulted in the formation of P-OH groups. In no case, phosphomolybdate complexes were formed.

The effects of modification of the SBA-16 surface with variable amounts of phosphorus on the acidity of oxide catalyst precursors and sulfided catalysts were investigated by the NH_3-TPD technique by Guzman et al. [25]. It was found that, regardless of the pretreatment, all P-containing samples show larger acidity than the P-free counterpart. The formation of $P(OH)_2$-O-Si bonds, or even free $OP(OH)_3$ entities interacting with the surface via H-bonding, during the support modification with phosphoric acid was proposed by Lewis et al. [46, 47] and Kawi et al. [31]. However, the theoretical calculation indicated that the formation of Si-O-P bonds could be excluded because they are very unstable [48].

The lowering of acidity after support grafting with a large amount of -PO_3H_3 groups observed by Guzman et al. [25] for NiMoW/P1.0/SBA-16 catalyst loaded with 1.0 wt.% of P_2O_3 (*vide infra* **Figure 4**) might indicate that a critical concentration of phosphoric acid is able to react with the surface hydroxyl groups. Above this critical concentration, H_3PO_4 reacts not only with the Si-OH groups but also with the P-OH groups to form polyphosphate species [46, 47]. Indeed, the XPS data analysis of those catalysts revealed that the binding energy of P_{2p} level of P-containing catalysts was much lower than those corresponding to P_2O_5 (134.0 vs. 135.2 eV) and similar (133.5 eV) to that reported for Si-MCM-41 grafted with -PO_3H_2 or -PO_2H groups [31].

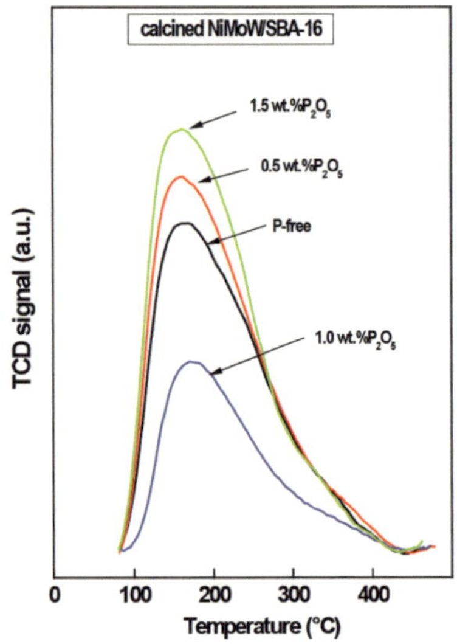

Figure 4. TPD-NH_3 profiles of oxide NiMoW/P(x)/SBA-16 catalyst precursors (x = 0.5–1.5 wt.%).

Moreover, it is known that the catalyst sulfidation led to a further increase in the catalyst acidity due to the formation of the -SH groups attached to the support surface [25]. However, among the catalysts loaded with different P contents, the sulfided NiMoW/P/SBA-16 catalysts with 1 wt.% P showed greater total acidity than its P-free counterpart, which may explain in part its best catalytic performance. On the contrary, the lower acidity of NiMoW/P/SBA-16 sulfide catalyst with a low P loading (0.6 wt.%) with respect to the P-free catalyst was explained in terms of the limited accessibility of ammonia to some acid sites due to the presence of small metal sulfide particles located on the substrate pores.

The increase of the catalyst acidity by support modification with phosphorus was also confirmed for NiMo/P/HMS-Al systems by Zepeda et al. [12, 22]. Their P-containing samples exhibited a greater acidity than their P-free counterpart, as confirmed by TPD-NH_3 study of oxide precursors and by DRIFT and FT-IR studies of adsorbed pyridine on the fresh sulfided catalysts.

8. Influence of phosphorus on the catalytic performance

The influence of phosphorus on the catalytic performance of transition metal sulfides was investigated in different hydrotreating reactions such as HDS, HDN, HDS+HDN, HDO, and HYD is shown in **Table 1**. Some reports showed that phosphorus acts as a promoter, while in many others, it was found to act as an inhibitor. Indeed, the most widely tested reaction to probe the performance of P-loaded silica was the HDS of dibenzothiophene (DBT). This is because DBT is a typical sulfur-containing molecule present in the petroleum fraction of high-boiling point or coal-derived liquids. Moreover, it is expected that DBT molecule did not show any diffusion limitation to access the porous structure of the mesoporous siliceous materials because this molecule covers a surface area of 8.0 × 12.2 Å². As compared to the DBT molecule, it is more difficult to desulfurize the alkyl-substituted DBTs, especially when both alkyl groups are in the 4- and 6-positions. Although the exact origin of the difficulty of converting alkyl-substituted DBTs using conventional alumina-supported Co(Ni)Mo catalysts still remains unclear, current studies associate this with (i) steric hindrance of the C-S bond scission in the adsorbed sulfur compound, (ii) the availability of only one H atom in the elimination step, and (iii) the influence of the alkyl group on the acidity of the H atom. The activity results for the HDS of 4,6-DMDBT confirmed that this reaction over CoMo/P/HMS-Ti sulfide catalysts proceeds via dealkylation and isomerization reaction pathways [13, 14]. The simplified reaction scheme for the HDS of 4,6-DMDBT reaction over those catalysts is shown in **Scheme 1**. In the HDS of DBT reaction, the DBT transformation occurs through parallel direct desulfurization (DDS) and hydrogenation (HYD) pathways: DDS route of DBT transformation leads to the formation of biphenyl (BP) via hydrogenolysis, whereas the HYD of one and two aromatic rings of DBT leads to the formation of tetrahydrodibenzothiophene (THDBT) and cyclohexylbenzene (CHB), respectively (see **Scheme 1**).

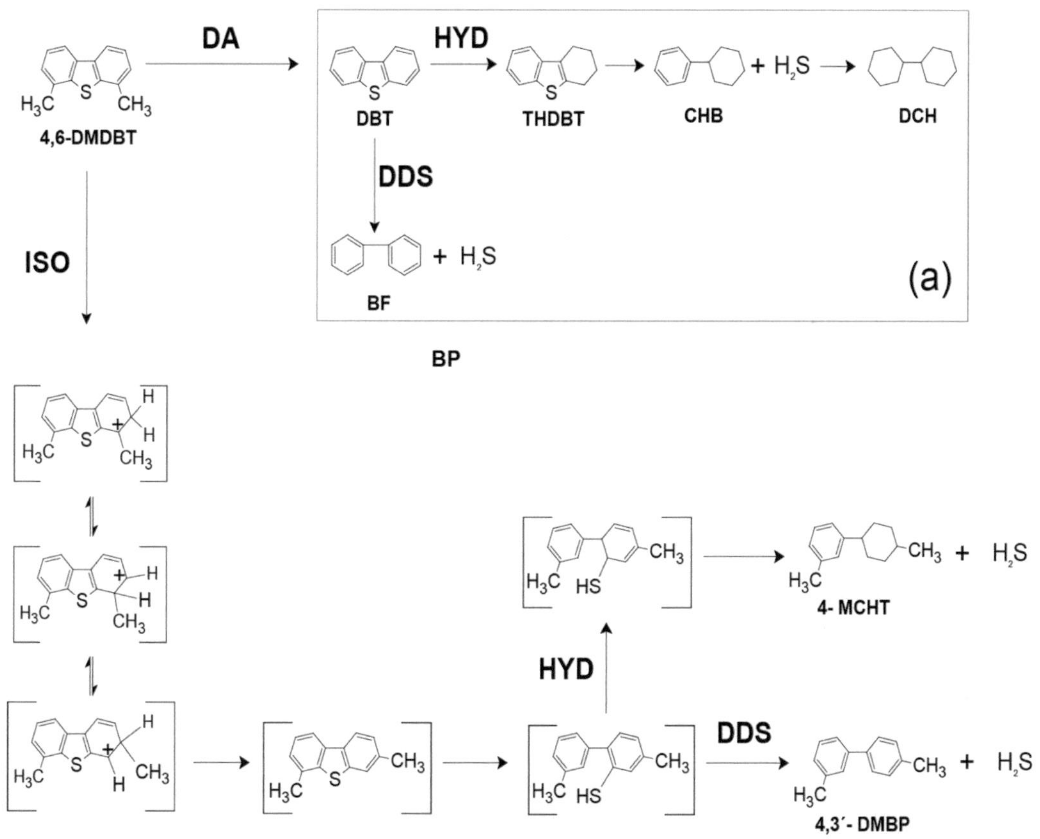

Scheme 1. Simplified reaction's schemes for the HDS of 4,6-DMDBT molecule based on the work of Nava et al. [17] (adapted with permission from Elsevier).

Catalyst[a]	Reactant	Reaction conditions	Reference
Phosphorus promotion effect			
$Ni/P/SiO_2$	Thiophene	Flow reactor, $T = 673$ K; $P = 1$ bar	Korányi [10]
NiMo/P/MCM-41	DBT, 4,6-DMDBT	Batch reactor, $T = 573$ K, $P = 7.3$ MPa	Herrera et al. [11]
CoMoP/Al-MCM-41	DBT	Batch reactor, $T = 593$ K, $P_{H2} = 5.6$ MPa	Hernández et al. [10]
Co/Mo/P/HMS	DBT	Batch reactor, $T = 623$ K, $P = 3.1$ MPa	Nava et al. [15, 17]
CoMo/P/HMS	DBT	Batch reactor, $T = 623$ K, $P = 3.1$ MPa	Nava et al. [15]
CoMo/P/HMS	4,6-DMDBT	Flow reactor, $T = 583$ K, $P = 5.0$ MPa	Nava et al. [17]
CoMo/P/HMS-Ti	DBT	Batch reactor, $T = 593$ K, $P = 5.5$ MPa	Pawelec et al. [13]
CoMo/P/HMS-Ti; Co/Mo/P/HMS-Ti	4,6-DMDBT	Batch and flow reactors, $T = 598$ K, $P = 5.5$ MPa	Pawelec et al., [14]
CoMo/P/HMS-Ti	DBT	Batch reactor, $T = 593$ K, $P = 5.5$ MPa	Zepeda et al. [21]
NiMo/P/HMS-Al	Thiophene, 4,6-DMDBT	Batch reactor, $T = 593$ K, $P = 5.5$ MPa	Zepeda et al., [22]
NiMo/P/HMS-Al	DBT	Batch reactor, T=593 K, P=5.5 MPa	Zepeda et al. [12]
NiMo/P/HMS-Al	Carbazole	Batch reactor, $T = 593$ K, $P = 5.5$ MPa	Zepeda et al. [12]
NiMo/P/HMS-Al	DBT+ carbazole	Batch reactor, $T = 593$ K, $P = 5.5$ MPa	Zepeda et al. [12]
NiMoW/P/SBA-16	DBT, 4,6-DMDBT	Batch reactor, $T = 593$ K, $P = 5.0$ MPa	Guzman et al. [25]
Phosphorus inhibition effect			
Co/Mo/P/HMS	DBT	Batch reactor, $T = 623$ K, $P = 3.1$ MPa	Nava et al. [15, 16]
NiMo/P/HMS-Ti; NiW/P/HMS-Ti	Naphthalene	Batch reactor, $T = 598$ K, $P = 7.7$ MPa	Halachev et al. [49]
CoMoW/P/SBA-15	DBT	Flow reactor, $T = 623$ K, P=3.1 MPa	Huirache-Acuña et al. [24]
CoMoW/P/SBA-15	Anisole	Flow reactor, $T = 583$ K, $P = 3.0$ MPa	Loricera et al. [26]
CoMoW/P/SBA-16	DBT	Flow reactor, $T = 623$ K, $P = 3.1$ MPa	Huirache-Acuña et al. [24]
CoMoW/P/SBA-16	Anisole	Flow reactor, $T = 583$ K, $P = 3.0$ MPa	Loricera et al. [26]

[a] Co/Mo (Ni/Mo): the catalysts prepared by successive impregnation; CoMo (NiMo): the catalysts prepared by simultaneous impregnation; CoMoP: simultaneous impregnation with metal salt precursors and H_3PO_4.

Table 1. Overview of the influence of phosphorus on the catalytic performance.

8.1. Effect of phosphorus concentration

A study reported by Nava et al. [50] clearly showed that the addition of phosphorus during preparation of unsupported NiMoW catalysts has detrimental effects on both textural and performance characteristics. The authors attributed the loss of activity to the loss of the Ni promotional effect resulting from the stronger interaction of P with Mo and W than with Ni. Moreover, the HRTEM results suggested that the proportion of $Mo(W)S_2$ slabs was drastically reduced after phosphorus addition. Contrary to the unsupported NiMoW, both inhibition and promotional effects were observed for the hydrotreating catalysts supported on the OMS substrates. An overview of the phosphorus effect for hydrotreating reactions is compiled in **Table 1**. It can be emphasized that most of the studies reporting the promotional effect of P were concentrated on the HDS reactions of transition metal sulfides.

The positive effect of P loading on the HDS activity of bifunctional NiMo catalysts supported on MCM-41 substrates was reported by Herrera et al. [11]. The surface of the MCM-41 substrate modified with variable amounts of phosphorus (0–5 wt.% P_2O_5) was prepared by post-synthesis grafting. In good agreement with the study by Guzman et al. [25], it was found that the modification of MCM-41 with phosphorus results in high-performance NiMo catalysts, especially for the HDS of DBT reaction. Regardless of the reaction, the promotion of the hydrogenation pathway of HDS reaction occurs. This was explained in terms of the changes in the morphology of MoS_2 active phase induced by phosphorus. Surprisingly, the P-containing catalysts exhibited a higher catalytic activity in 4,6-DMDBT HDS than in DBT HDS reaction, with all P-containing catalysts being more active in the 4,6-DMDBT HDS reaction than a commercial $NiMo/Al_2O_3$ catalyst. Maximum catalytic activity in HDS of 4,6-DMDBT reaction was observed for the catalyst supported on MCM-41 modified with 1 wt.% P_2O_5.

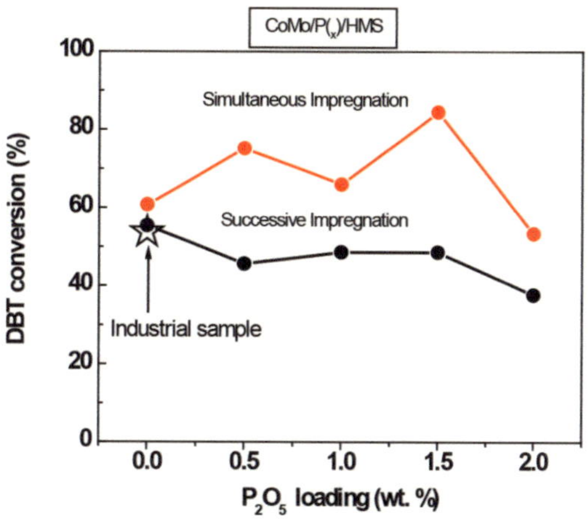

Figure 5. Influence of P loading and catalyst preparation method (successive impregnation vs. simultaneous impregnation) on the activity of CoMo/HMS sulfide catalysts in HDS of DBT. As reference, the DBT conversion data of an industrial sample is included (adapted from Nava et al. [15] with permission from Elsevier).

The influence of P loading on the HDS activity of P-containing CoMo/P/HMS sulfide catalysts prepared by sequential and successive impregnation was investigated by Nava et al. [15]. Their P-containing substrates were prepared by incipient wetness impregnation of the parent HMS support with an aqueous solution of H_3PO_4 having an appropriate concentration to obtain P/HMS supports with P_2O_5 loading in the range of 0.5–2.0 wt.%. **Figure 5** shows that the P effect is not linear and strongly depends on the method of catalyst preparation. The best HDS activity was shown by the catalyst loaded with 1.5 wt.% of P_2O_5 which was prepared by simultaneous impregnation.

The effect of high P loading (2.0 wt.% of P_2O_5) on the surface of HMS-based catalysts was different for the HDS of DBT [15] and the HDS of 4,6-DMDBT reactions [17]. Thus, contrary to HDS of DBT, the same high P loading (2.0 wt.% of P_2O_5) favored the catalytic behavior of CoMo/P/HMS sulfide catalysts in the HDS of 4,6-DMDBT reaction. The activity enhancement was associated with the larger sulfidation degree of cobalt species as confirmed by XPS analysis. However, both phosphorus promotion in the HDS of DBT [15] and inhibition in the HDS of 4,6-DMDBT [17] were observed for the CoMoS/P(x)/HMS catalysts prepared by simultaneous impregnation (**Figure 6**). Thus, it is obvious that the effect of modification of HMS substrate with P on the catalyst activity is contradictory. The drop in activity in the HDS of DBT reaction over CoMo/P-HMS sulfide catalyst (2.0 wt.% of P) was explained in terms of the much lower specific surface area and the presence of a large surface concentration of unsulfided Co^{2+} species on the P-containing sample [16].

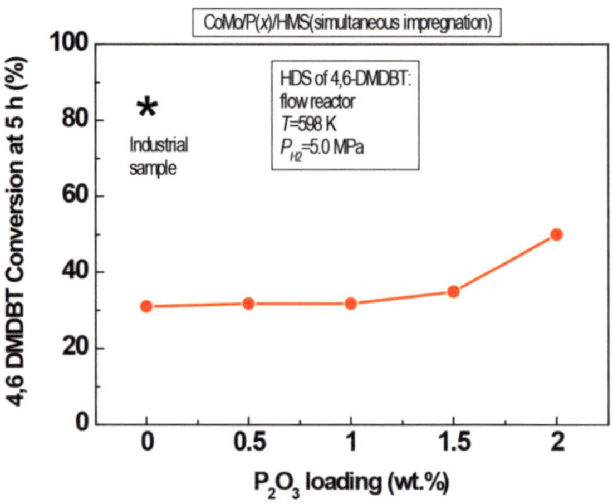

Figure 6. Influence of P loading on the activity of CoMo/HMS sulfide catalysts in the HDS of 4,6-DMDBT. The catalysts were prepared by simultaneous impregnation. As reference, the 4,6-DMDBT conversion data of an industrial sample is included (adapted from Nava et al. [17] with permission from Elsevier).

The promotional P effect was also observed for P/SBA-16-supported NiMoW catalysts when used in both HDS of DBT and 4,6-DMDBT reactions [25]. For both reactions, the catalyst activity plotted against P loading exhibited a volcano-type curve. The volcano-curve activity trend reported for NiMoW/P/SBA-16 sulfide catalysts was attributed to the promotional effect of P leading to an increase of the dispersion (from HRTEM). However, the catalyst loaded with the largest amount of phosphorus (1.6 wt.%) exhibited a drop of HDS activity with respect to its P-free counterpart. The low HDS activity of the P-free counterpart with respect to the P-containing NiMoW/SBA-16 sulfide catalysts was explained in terms of the formation of the "onion-type" $Mo(W)S_2$ phases (*vide supra* **Figure 3**), in good agreement with the study by Huang et al. [51]. This is because the "onion-type" phase exhibits the closed shell structure of MoS_2 which displays a low amount of HDS active edge sites. Regardless of the P content, DBT and 4,6-DMDBT transformation over NiMoW/P/SBA-16 catalysts occurred through two parallel routes of hydrogenation (HYD) and direct desulfurization (DDS) reaction routes [25].

Figure 7 shows the influence of P loading on the HYD/DDS selectivity ratio in this reaction over NiMoW/P/SBA-16 sulfide catalysts. As seen in this figure, the catalysts modified with P exhibited the enhancement of HYD route of the DBT HDS reaction, in good agreement with those observed for CoMoW/SBA-16 by Huirache-Acuña et al. [24].

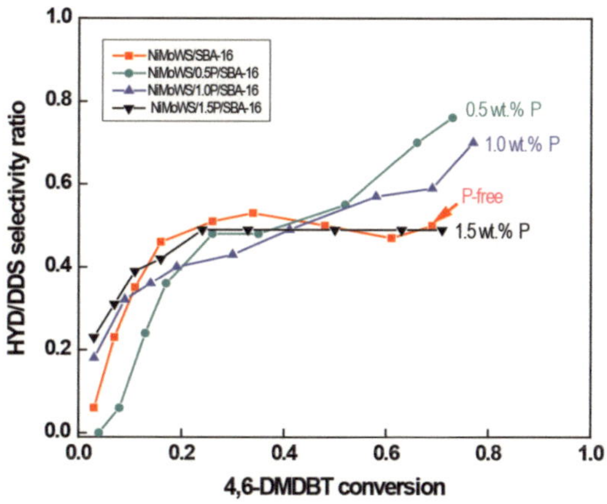

Figure 7. Influence of P loading on the HYD/DDS selectivity ratio in the HDS of 4,6-DMDBT over NiMoW/P/SBA-16 sulfide catalysts (adapted from Guzman et al. [25] with permission from Elsevier). Reaction conditions were: a batch reactor, $T = 320°C$, $P = 5.0$ MPa, and $t = 0.5$ h.

In order to promote the migration of the methyl groups in the aromatic ring of 4,6-DMDBT, attempts were made to enhance the acid properties of support by the substitution of Si^{4+} atoms by Al^{3+} or Ti^{4+} ions in the pore walls of mesoporous silicas. The effect of support modification with both P and Al^{3+} (or Ti^{4+}) on the activity of OMS-based catalysts has been studied by several authors [13, 14, 21, 22, 34]. In all these studies, P was incorporated by impregnation of the parent substrate with H_3PO_4. Zepeda et al. [21] investigated the catalytic response of CoMo/P/HMS-Ti sulfide catalysts in the HDS of DBT reaction carried out in a batch reactor at 320°C and total H_2 pressure of 5.5 MPa. It was found that HDS activity strongly increased upon P loading up to 0.64 wt.% of P_2O_5. The most active catalyst was 3.7 times more active than a commercial reference one and 2.4 times more active than its P-free counterpart. From the catalyst characterization, this was explained as due to specific electronic properties of the active phases and the largest Mo surface exposure on the support surface (from XPS).

The effect of phosphorus loading was also studied for sulfided NiMo/P/HMS-Al catalysts by Zepeda et al. [22]. For both thiophene and 4,6-DMDBT HDS reactions, the activity of P-loaded sulfide catalysts against P_2O_5 loading follow a volcano-shaped curve. The highest activity was exhibited by the catalyst doped with an optimized amount of P (1.0 wt.% of P_2O_5). This was explained in terms of the appropriate balance of active phase dispersion and largest hydrogenation ability of the catalysts. Indeed, a linear dependence was observed between catalyst activity and the active phase dispersion derived from H_2 chemisorption. The detrimental effect of a large P loading (1.5 and 2.0 wt.% of P_2O_5) on the HDS activity of NiMo/P/HMS-Al catalysts was attributed to a large decrease in their specific surface area with respect to P-free catalysts [22]. Moreover, the decrease in the active phase dispersion of the catalysts was consistent with a more difficult sulfidation of the support-attached tetrahedral Mo species, as confirmed by combined DRS UV-vis and temperature-programmed sulfidation (TPS) experiments of the oxide precursors.

8.2. Effect of the catalyst preparation method

The most widely used preparation method of the Co(Ni)-promoted Mo(W) catalysts is the wetness impregnation of P/OMS supports with metal salt precursors. The direct preparation based on the dissolution of ammonium heptamolybdate in the presence of phosphoric acid has also been reported. In such cases, phosphorus acid is first added to an aqueous solution of ammonium heptamolybdate before adding nickel (cobalt) [50].

The Co(Ni)-promoted Mo-based catalysts are usually prepared by simultaneous or sequential impregnation of the P/OMS substrates with Co(Ni) and molybdenum salts solutions. The influence of both methods on the HDS activity of phosphate-containing CoMo/P/HMS sulfide was studied by Nava et al. [15]. The effect of the catalyst preparation method on the HDS activity of that catalysts is shown in **Figure 5** (*vide supra*).

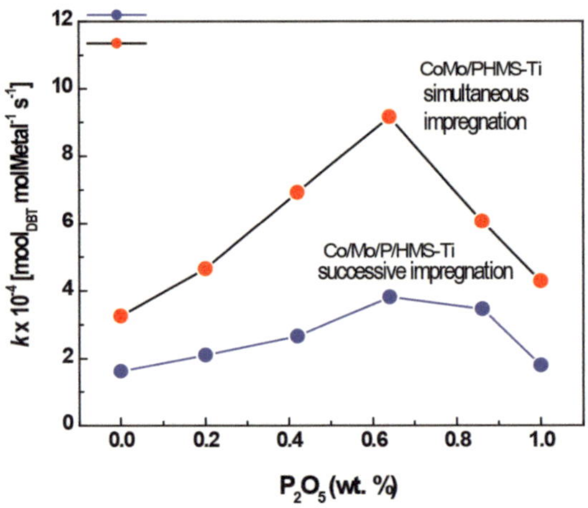

Figure 8. Influence of the catalyst preparation method and support modification with P on the pseudo-first reaction rate constants in the HDS of DBT over sulfided Co/Mo/P(x)/HMS-Ti and CoMo/P(x)/HMS-Ti catalysts (adapted from Pawelec et al. [13] with permission from Elsevier).

The influence of the catalyst preparation method (successive vs. simultaneous impregnation) on the HDS activity of P-containing CoMo/HMS-Ti sulfide catalysts in the HDS of DBT reaction was studied by Pawelec et al. [13]. **Figure 8** shows the reaction rate constants calculated from the DBT conversion at reaction time of 4 h. As seen in this figure, the simultaneous P/HMS-Ti support impregnation with Co and Mo precursors led to a larger enhancement of HDS activity than that prepared by successive impregnation. For both catalyst series, the promotion of P was not linear and the optimum P loading was found to be close to 0.64 wt.% (P_2O_5 content). Importantly, it was found that the catalyst supported on the P/HMS-Ti substrate having an

http://dx.doi.org/10.5772/64181

optimized P loading (0.64 wt.% of P_2O_5), which was prepared by simultaneous impregnation, exhibited a larger HDS activity than an industrial catalyst having the same formulation.

The effect of the catalyst preparation method and support modification with P on the reaction mechanism of the 4,6-DMDBT transformation over CoMo/P/HMS-Ti sulfide catalysts was studied by Pawelec et al. [14]. This is contrary to the NiMo/P/HMS-Ti sulfide catalysts which showed the 4,6-DMDBT transformation via HYD and DDS reaction routes [22]. The post-synthesis modification of HMS-Ti with phosphorus led to an enhancement of dealkylation and isomerization routes of 4,6-DMDBT transformation over the CoMo/P/HMS-Ti sulfide catalysts [14].

8.3. Effect of phosphorus addition on coke formation

By means of the TPO/TG analysis of the spent P-free and P-containing CoMo/HMS catalysts tested in HDS of DBT, Nava et al. [17] observed that P-containing catalysts were more susceptible to coking and produced more hydrogen-rich coke than the P-free counterpart. Besides the larger coke formation, the CoMo/HMS catalyst with the largest P loading (2.0 wt.% of P_2O_3) exhibited the best performance in the HDS reaction. From the catalyst characterization, this was ascribed to a proper balance between active phase dispersion and the lowest deactivation among the catalysts studied. This is because, besides the coke formation, other factors might contribute to catalyst deactivation. Thus, the opposite trend observed for the catalyst stability and the amount of coke formed strongly suggested that the presence of phosphate species on the support surface could avoid particles sintering during on-stream conditions. Moreover, the presence of phosphate species on the support surface helps maintain a large surface area of the supported catalysts, as observed for the alumina-supported ones [46, 47]. This means that an optimum loading of phosphorus may increase the longevity of the catalysts. However, the introduction of high amounts of phosphate led to an increase in the catalyst acidity, as confirmed for the NiMo/Al-HMS-P catalysts by Zepeda et al. [22], leading to promotion of coke formation. In this context, Stanislaus et al. [52] demonstrated that the strong acid sites of the alumina substrate, which are important in coking, are progressively poisoned by phosphate addition. However, the introduction of a high amount of phosphate created medium strength acid sites, and in turn, the total acidity of support increases. As a consequence, one might expect that the coking might also increase.

The influence of the support morphology (SBA-15 vs. SBA-16) and their modification with phosphate on the coking behavior of sulfided CoMoW catalysts was investigated by Huirache-Acuña et al. [24]. In general, it was reported that the spent SBA-16-based catalysts showed a lower coke formation in the hydrodesulfurization (HDS) of dibenzothiophene (DBT) than their SBA-15-based counterparts. This is because the phosphate incorporation to the SBA-16 support led to a decrease in coke formation in the final catalysts, as opposite to the behavior of the SBA-15-based counterparts. Interestingly, it was found that the phosphate addition to SBA-15 substrate led to an increase in coke formation, whereas for the SBA-16-based catalysts, the incorporation of a large amount of phosphate (1.5 and 2 wt.% of P) followed an opposite trend. Since a correlation was found between the amount of weak and medium strength acid sites and the amount of coke, this might indicate that coke formation on this sample occurs on the acid sites that are located in the close vicinity of CUS sits. Multipoint centers for coke adsorption have been suggested as the sites that are first poisoned by coke [53].

Summarizing, regardless of the support morphology, the optimization of P loading seems to be an important parameter in the design of the active hydrotreating catalysts supported on the ordered mesoporous siliceous substrates. This is because phosphorus content usually affects the active phase dispersion, the catalyst acidity, the metal-support interaction, and the number of active sites. It was found that the support modification with P affected the following aspects of the final catalyst: (i) its structural and textural properties (S_{BET}); (ii) its acidity, and (iii) the formation of different phases in the oxide precursors. However, the addition of phosphorus to mesoporous siliceous materials has a nonlinear effect on the catalytic activity because the incorporation of a large amount of phosphorus (>0.1 wt.%) led to a decrease in the HDS activity of hydrotreating catalysts. This is probably due to a weakening of the metal-support interaction leading to a lower dispersion of the metal oxide phases, similar to that observed for alumina-supported hydrotreating catalysts [3].

9. Conclusions

On the basis of the present revision on the effect of modification of mesoporous silica materials by -PO_3H_2 groups for the activity of sulfides catalysts in different hydrotreating reactions, the following conclusions can be drawn:

i. The effect of the modification of the mesoporous silicas with phosphorus strongly depends on the P loading and method of its incorporation.

ii. The amount of phosphorus should be optimized and the best activity results are obtained for the catalysts modified with 1 wt.% of P.

iii. Support functionalization with -PO_3H_2 groups by post-synthesis method increases the substrate acidity without deterioration of its morphology.

iv. The catalytic behavior of sulfided catalysts depends strongly on the characteristics of their oxide precursors. Thus, a linear correlation was found between the presence of irregular (bankrupt) oxidized Mo particles in the oxide precursors and the catalytic response.

v. The deposition of oxide Co and Mo precursors by simultaneous impregnation led to more active HDS catalysts than their deposition on the support surface by successive impregnation method. This is because the former method led to a larger BET-specific surface area, a larger amount of Co^{2+} ions having octahedral symmetry, and a larger amount of Mo species having $Mo{=}O_T$ bonds.

vi. Irrespective of the catalyst preparation method, the presence of phosphate species on the surface of HMS-Ti substrates has a beneficial effect on the HDS activity of the catalysts because of the enhanced formation of octahedral Co^{2+} species and irregular Mo^{6+} particles in the oxide precursors.

vii. Compared with classical γ-Al_2O_3, the HMS support modified with both Ti (Si/Ti molar ratio of 40) and phosphate shows appropriate textural characteristics to be used as support for HDS catalysts.

Acknowledgements

Dr. R. Huirache-Acuña acknowledges the financial support of CONACYT 182191 and CIC-UMSNH 2016-2017 projects, whereas Dr. B. Pawelec wishes to express her gratitude for financial support to Comunidad de Madrid (Project CAM S2013/MAE-2882).

Author details

Rafael Huirache-Acuña[1*], Eric M. Rivera-Muñoz[2], Trino A. Zepeda[3], Rufino Nava[4] and Barbara Pawelec[5]

*Address all correspondence to: rafael_huirache@yahoo.it

1 Chemical Engineering Faculty, Michoacan University of Saint Nicholas of Hidalgo, Morelia, Michoacán, México

2 Center of Applied Physics and Advanced Technology, National Autonomous University of Mexico, Querétaro, Qro., México

3 Center of Nanoscience and Nanotechnology, National Autonomous University of Mexico, Ensenada, BC, México

4 Engineering Faculty, Autonomous University of Querétaro, Querétaro, Qro., México

5 Institute of Catalysis and Petroleochemistry of the Spanish National Research Council, Cantoblanco, Madrid, Spain

References

[1] Vasudevan P.T., Fierro J.L.G. *Catalysis Review-Science Engineering*. 1996; 38: 161–188.

[2] Breysse M., Afanasiev P, Geantet C., Vrinat M. *Catalysis Today*. 2003; 86: 5–16. DOI: 10.1016/S0920-5861/(03)00400-0.

[3] Iwamoto R., Grimblot J. *Advances in Catalysis*. 2000; 44: 417–503. DOI: 10.1016/S0360-0564(08)60516-7.

[4] Lai, W.F., Hamilton M.A., McCarthy S.J. WO2013059176 A1. 2013.

[5] Fierro J.L.G, López Agudo A., Esquivel N., López Cordero R. *Applied Catalysis* 1989; 48:353–363.

[6] Robinson, W.R.A.M., van Gestel J.N.M., Korányi T.I., Eijsbouts, S., van der Kraan A.M., van Veen, J.A.R., de Beer V.H.J. *Journal of Catalysis*. 1996; 161:539–550. No. 0021-9517/96. No. 0166-9834.

[7] Topsøe H., Clausen B.S., Massoth F.E. in: J.R. Anderson, M. Boudart (Eds.) Hydrotreating Catalysts. Science and Technology, Springer-Verlang, Berlin, 1996. Vol. 11.

[8] Gishti K., Iannibello A., Marengo S., Morelli G., Tittarelli P. *Applied Catalysis*. 1984; 12: 381–393. 0166-9834/84.

[9] Van der Bij H.E., Weckhuysen B.M. *Chemical Society Reviews*. 2015; 44: 7406–7428. DOI: 10.1039/C5CS00109A.

[10] Korányi T.I. *Applied Catalysis A: General*. 2003; 239: 253–267.PII: S092-860X(02)00390-3.

[11] Herrera J.M., Reyes J., Roquero P., Klimova T. *Microporous and Mesoporous Materials*, 2005; 83: 283–291.DOI: 10.1016/j.micro.meso.2005.05.010.

[12] Zepeda T.A., Pawelec B., Obeso-Estrella R., Díaz de León J.N., Fuentes S., Alonso-Nuñez G., Fierro J.L.G. *Applied Catalysis B: Environmental* 2016; 180: 569–579. DOI: 10.1016/j.apcatb.2015.07.013.

[13] Pawelec B., Halachev T., Olivas A., Zepeda T.A. *Applied Catalysis A: General* 2008; 348: 30–41. DOI: 10.1016/j.acata.2008.06.014.

[14] Pawelec B., Fierro J.L.G., Montesinos A., Zepeda T.A. *Applied Catalysis B: Environmental* 2008; 80: 1–14. DOI: 10.1016/j.apcatb.2007.10.039.

[15] Nava R., Morales J., Alonso G., Ornelas C., Pawelec B., Fierro J.L.G. *Applied Catalysis A: General* 2007; 321: 58–70. DOI: 10.1016/j.acata.2007.01.038.

[16] Nava R., Pawelec B., Morales J., Ortega R.A., Fierro J.L.G. *Microporous and Mesoporous Materials*, 2009; 118: 189–201. DOI: 10.1016/j-micromeso.2008.08.045.

[17] Nava R., Infantes-Molina A., Castaño P., Guil-López R., Pawelec B. *Fuel*, 2011 90: 2726–2737. DOI: 10.1016/j.fuel.2011.03.049.

[18] Kresge J.S., Leonowicz M.E., Roth W.J., Vartuli J.C., Beck J.S. *Nature* 1992; 359: 710–712. DOI: 10.1038/359710a0.

[19] Tanev P.T., Chibwe M., Pinnavaia T.J. *Nature*, 1994; 368: 321–323. DOI: 10.1038/368321a0.

[20] Tanev P.T., Pinnavaia T.J. *Chemistry of Materials* 1996; 8: 2068–2079. DOI: 10.1021/cm950549a.

[21] Zepeda T.A., Pawelec B., Fierro J.L.G., Montesinos A., Olivas A., Fuentes S., Halachev T. *Microporous and Mesoporous Materials* 2008; 111: 493–506. DOI: 10.1016/j.micromeso.2007.08.027.

[22] Zepeda T.A., Infantes-Molina A., Díaz de León J.N., Fuentes S., Alonso-Núñez G., Torres-Otañez G., Pawelec B. *Applied Catalysis A: General* 2014; 484, 108–121. DOI: 10.1016/j.apcatb.2014.06.033.

[23] Zhao D., Feng J., Huo Q., Melosh N., Fredrickson G.H., Chmelka B.F., Stucky G.D. *Science* 1998; 279: 548–552.

[24] Huirache-Acuña R., Pawelec B., Rivera-Muñoz E., Nava R., Espino J., Fierro J.L.G. *Applied Catalysis B: Environmental*, 2009; 92: 168–184. DOI: 10.1016/j.apcatb.2009.07.012.

[25] Guzmán M.A., Huirache-Acuña R., Loricera C.V., Hernández J.R., Díaz de León J.N., de los Reyes J.A., Pawelec B. *Fuel* 2013; 103: 321–333. DOI: 10.1016/j.fuel.2012.07.005

[26] Loricera C.V., Pawelec B., Infantes-Molina A., Alvarez-Galván M.C., Huirache-Acuña R., Nava R., Fierro J.L.G. *Catalysis Today* 2011; 172: 103–110. DOI: 10.10916/j.cattod.2011.02.037.

[27] Sakamoto Y, Keneda M., Teresaki O., Zhao D.Y., Kim J.M., Stucky G., Shin H.J., Ryoo R. *Nature* 2000; 408: 449–453. DOI:10.1038/35044040.

[28] Pitchumani R., Li W., Coppens M.-O. *Catalysis Today* 2005; 105: 618–622. DOI: 10.1016/j.cattod.2005.06.002.

[29] Tagaya M., Kobayashi K., Nishikawa M. *Materials Letters* 2016; 164:651–654. DOI: 10.1016./j.matlet.2015.11.070.

[30] Vallet-Regí M., Izquierdo-Barba I., Rámila A., Pérez-Pariente J., Babonneau F., González-Calbet *Solid State Science.* 2005; 7: 233–237. DOI: 10.1016/j.solidstatesciences.2004.10.038.

[31] Kawi S., Shen S.C., Chew P.L. *Journal of Materials Chemistry*. 2002; 12, 1582–1586. DOI: 10.1039/b107795n.

[32] González M.D., Salagre P., Mokaya R., Cesteros Y. *Catalysis Today* 2014; 227: 171-178. DOI: 10.1016/j.cattod.2013.10.029.

[33] Gontier, S., Tuel A. *Zeolites* 1995; 15:601–610. DOI:10.1016/0144-2449(95)00028-5.

[34] Hernández A., Escobar J., Pacheco, J.G., de los Reyes J.A., Barrera, M.C. *Journal of the Mexican Chemical Society*, 2004; 48: 260–268.

[35] Van der Bij H.E., Cicmil D., Wang J., Meirer F., de Groot F.M.F., Weckhuysen B.M. *Journal of American Chemical Society* 2014; 136(51): 17774–17787. doi: 10.1021/ja508545m.

[36] Lezcano-Gonzalez I., Deka U., van der Bij H.E., Paalanen P., Arstad B., Weckhuysen B.M., Beale A.M. *Applied Catalysis B: Environmental* 2014; 154-155:339–349. DOI: 10.1016/j.apcatb.2014.02.037.

[37] Van der Bij H.E., Weckhuysen B.M. *Physical Chemistry and Chemical Physics* 2014; 16: 9892–9903. DOI: 10.1039/c3cp54791d.

[38] Van der Bij H.E., Aramburo L.R., Arstad B., Dynes J.J., Wang J., Weckhuysen B.M. *ChemPhysChem*. 2014; 15: 283–292. DOI: 10.1002/cphc.201300910.

[39] Shon J.K., Yuan X., Ko C.H., Ik Lee H., Thakur S.S., Kang M., Kang M.S., Li D., Kim J.N., Kim J.M. *Journal of Industrial Engineering Chemistry*. 2007; 13(7): 1201–1207.

[40] El Mourabit S., Guillot M., Toquer G., Cambedouzou J., Goettmann F., Grandjean A. *RSC Advances* 2012; 2: 10916–10924. DOI: 10.1039/c2ra21569a.

[41] Xiong L., Yang Y., Shi J., Nogami M. *Microporous and Mesoporous Materials*. 2008; 111: 343–349. DOI: 10.1016/j.micromeso.2007.08.016.

[42] Spanos N., Lycourghiotis A. *Journal of Colloid Interface Science*. 1995; 171: 306–318. DOI: 10.1006/jcis.1995.1185.

[43] Oyama S.T. *Journal of Catalysis* 2003; 216: 343–352. DOI: 10.1016/S0021-9517(02)00069-6.

[44] Huirache-Acuña R., Nava R., Peza-Ledesma C.L., Lara-Romero J., Alonso-Núñez G., Pawelec B., Rivera-Muñoz E.M. *Materials* 2013; 6:4139–4167.

[45] Wang Y., Shen B., Hao K., Li J., Wang L.; Feng B., Guo Q. Catalysis Communications 2012; 25: 59–63. DOI: 10.1016/j.catcom.2012.04.004.

[46] Lewis J.M., Kydd R.A. *Journal of Catalysis*, 1991; 132: 465–471. DOI: 10.1016/0021-9517(91)90163-X.

[47] Lewis J.M., Kydd R.A., Boorman P.M., van Rhyn P.H. *Applied Catalysis A: General*, 1992; 84: 103–121. DOI: 10.1016/0926-860X(92)80110-X.

[48] Lű, R., Cao, Z., Wang, S. *THEOCHEM*, 2008; 865:1–7. DOI: 10.1016/j.theochem. 2008.06.009.

[49] Halachev T., Nava R., Dimitrov L. *Applied Catalysis A: General*, 1998; 169: 11–117. *PII* S0926-860X(97)00374-8.

[50] Nava H, Espino, J., Berhault G., Alonso-Nuñez G. *Applied Catalysis A: General*, 2006; 302: 177–184. DOI: 10.1016/j.apcata.2005.12.025

[51] Huang L., Li Q. *Chemistry Letters*, 1999; 28(8): 829–830. DOI: 10.1246/cl.1999.829

[52] Stanislaus A., Absi-Halabi M., Al-Doloma K. *Applied Catalysis* 1988; 39: 239–253. DOI: 10.1016/S0166-9834(00)80952-5.

[53] Ione K.G., Echeviskii G.V., Nosyreva G.N. *Journal of Catalysis*. 1984; 85: 287–294. No. 0021-951718.

Chapter 7

Microporous and Mesoporous Materials in Decontamination of Water Process

Rafael Alberto Fonseca-Correa,
Yesid Sneider Murillo-Acevedo,
Liliana Giraldo-Gutiérrez and
Juan Carlos Moreno-Piraján

Additional information is available at the end of the chapter

http://dx.doi.org/10.5772/64393

Abstract

Water pollution is the main threat that confronts humanity today. The daily human activity introduces different pollutants in hydric sources that modify water quality. This represents a high production of water that is not appropriate for different applications like human consumption. Among the water pollutants more common are dyes, herbicides, pesticides, alkanes, halo alkanes, aliphatic compounds, alcohols, carboxylic acids, aromatic compounds, detergents, surfactants, inorganic compounds (heavy metals), harmful gases, and pathogenic bacteria (bacteria and virus). These pollutants can be found in underground and surface water, and therefore can produce adverse effects on the environment and human health. These effluents can be treated with different procedures such as biological and chemical treatments, among others. However, in some cases, these processes are not appropriate for compliance legislation. Thus, adsorption with special materials such as carbon aerogels and advance oxidation processes such as photocatalysis are widely studied nowadays.

Keywords: aerogels, supercritical drying, SBA-15, photodegradation, photocatalysis

1. Introduction

The microporous materials have been used in the water decontamination process because of their adsorption capacity of a wide variety of contaminants in aqueous phase. The carbon aerogels have become a promising material in adsorption processes of ions. The textural properties have an important function in the adsorption process. However, the adsorption treatment is not sufficient and for this reason research on new processes is necessary in water decontamination.

Within the new processes in water treatment, it has been found that the advanced oxidation process, photocatalysis, is of great importance. The photocatalyst, TiO_2, is used more because of its low surface area, which is a very important factor in the photocatalytic activity. For this reason, the mesoporous materials of type SBA-15 support TiO_2.

Therefore, the processes such as adsorption with carbon aerogels and degradation of organic compounds with Ti-SBA-15 (photocatalyst) are used for water decontamination. The textural properties of these porous materials are different but they can be complementary in processes of water decontamination.

2. Microporous and mesoporous materials

2.1. Inorganic ions adsorption on carbon aerogel

In recent years, the synthesis of nanoporous solids has received much attention, and important methods to fabricate their pore structures have been proposed recently [1–5], such as carbon aerogels. The pore continuity of aerogels, sol-gel process has attracted much attention for the fabrication of porous materials [5]. Aerogels refer to dried gels with a very high relative pore volume [6]. Pekala succeeded in the preparation of carbon aerogels, which are carbonized from organic polymer aerogels in an inert atmosphere such as nitrogen [7, 8]. Due to large surface area and easily tunable nanoporosities, these materials have potential applications in various fields, such as hydrogen fuel storage, catalysis, supercapacitors, electrodes, and anodes in rechargeable lithium ion batteries [9]. In this chapter, we show a new application of these materials to remove ions from aqueous solution, for example, nickel and chromium.

The aerogels could be divided into monolith, powder, and film; and by considering the preparation method, aerogel could be made up of four types including aerogel, xerogel, cryogel, and other aerogel-related materials, hydrogels or alcohogels. While given the different microstructure, aerogel could be classified as microporous (<2 nm) aerogel, mesoporous (2–50 nm) aerogel, and mixed-porous aerogel.

The application design of the carbon aerogel is based on its properties, which rely on the microstructure. Therefore, it is very important to realize the microstructure control during the preparation. Commonly, the preparation process of the aerogel includes the following three key steps [10], as shown in **Figure 1**:

(I–II) *Solution-sol transition*: nanoscale sol particles are formed in the precursor solution spontaneously or catalyzed by the catalysts via hydrolysis and condensation reactions.

(III–IV) *Sol-gel transition* (*gelation*): the sol particles are cross-linked and hierarchically assembled into a wet gel with the coherent network.

(V–VI) *Gel-carbon aerogel transition* (*drying and carbonization*): the solvent inside the wet gel was replayed by the air without serious microstructure damage and obtained the carbonaceous structure from pyrolysis.

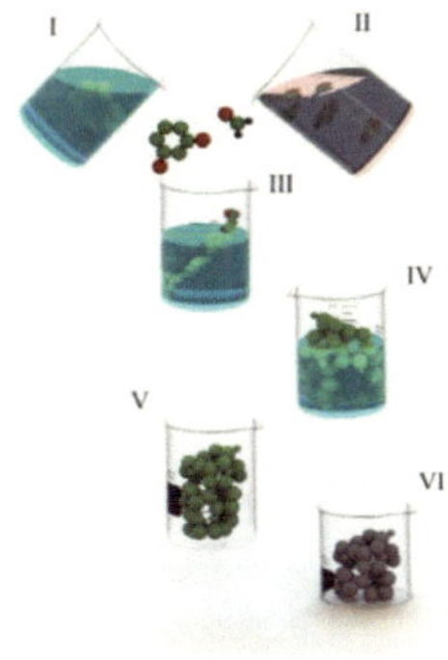

Figure 1. Synthesis of aerogels.

All the three steps could determine the microstructure of the aerogel and affect its properties and applications [11].

Carbon aerogel was prepared by the Pekala method via Na_2CO_3-catalyzed polycondensation of resorcinol with formaldehyde (RF aerogel) in an aqueous solution [12]. In fact, the random network of RF gel has been built by the homogeneous polymerization of resorcinol and formaldehyde in a large proportion of solvent (water) (**Figure 2**). The mixtures are stirred for a certain period of time and poured into cylindrical glass molds with lid.

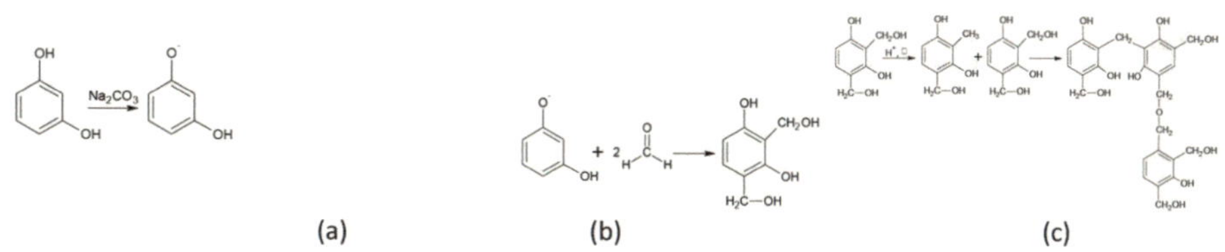

Figure 2. Reaction of polymerization. (a) Enolate ion formation, (b) formation of the monomer, and (c) obtaining cross-linked polymer network [12].

The resorcinol-formaldehyde wet gels were prepared using resorcinol (98% purity), formaldehyde (37% solution), Na_2CO_3 (99.9% purity), all from Aldrich, and deionized water. Resorcinol (0.29 mol) was dissolved in deionized water at a certain R/W ratio. Solution of formaldehyde was added to the resorcinol solution (R/F = 0.5) in vigorous stirring. Afterward, Na_2CO_3 in 0.1 M aqueous solution was added to the previous mixture at different ratios and R/C 100, 400, and 600 samples were obtained. The solutions were placed into tightly closed glass molds (7 cm length × 1 cm internal diameter) and cured: However, the rate of these reactions in the room temperature is very slow. Normally, a multiple-stage heating process is used to accelerate the gelation, 1 day at room temperature, 1 day at 50°C, and 3 days at 70°C.

Conventionally, the term "aerogels" has been use to designate gels dried under supercritical conditions. Supercritical drying mostly affects only the larger pores that involve network dimensions. The critical point parameters of CO_2, which is the most commonly used fluid, are as follows: T_c (°C) 31.0, P_c (bar) 74 [13]. The removal of water becomes essential in the case of supercritical drying (**Figure 3**) with CO_2 because of the insolubility between CO_2 and water, the aqueous solvent is replaced with an organic one (e.g., methanol, acetone, isopropanol, or amyl acetate) through a repetitive washing procedure. The resulting resorcinol-formaldehyde gels were washed with ethanol followed by acetone and dried with CO_2 in supercritical conditions (38–40°C, 120 bar) resulting in resorcinol-formaldehyde aerogels.

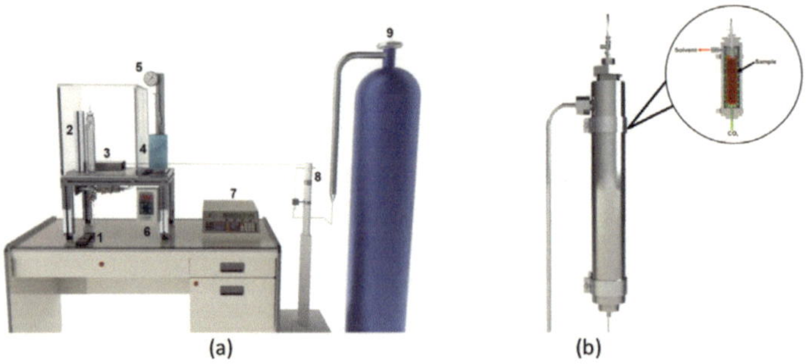

Figure 3. (a) Equipment for supercritical fluid used for drying the aerogels: (1) thermocouple, (2) extractor, (3) finned resistor and ventilator, (4) hot air bath, (5) condensation trap-pressure gauge, (6) temperature controller, (7) control unit of the syringe pump, (8) syringe pump, and (9) CO_2 cylinder and (b) extractor.

Pyrolysis of the organic polymer aerogels produces carbon aerogels, which are known to have uniform mesopores that depend on the agglomerate structures of uniform spherical carbon particles. The basic idea is to prepare carbonized RF (CRF) aerogels. The organic aerogels were pyrolyzed in N_2 atmosphere for 4 hours, or at 850°C samples, resulting in carbon aerogels. It is possible to obtain a CRF aerogel with ultrahigh-specific surface area by activating the carbon skeletons with carbon dioxide under high temperature (normally 800–1200°C) [14]. Carbon dioxide rather corrodes than activates the skeletons, creating more pores (mainly micropore).

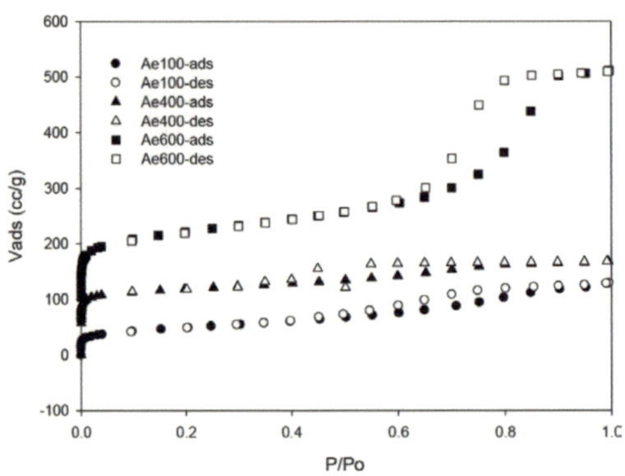

Figure 4. Adsorption and desorption isotherms of nitrogen on resorcinol-formaldehyde carbon aerogels at 77 K.

The aerogels are known to have outstanding characteristics (e.g., higher surface areas, total pore volumes, and sometimes electrochemical double-layer capacitances).

The structures of resorcinol-formaldehyde aerogels depend on the size and number of resorcinol-formaldehyde clusters generated during the polymerization in the synthesis process. Low-temperature nitrogen adsorption measurement is a common means for characterization of nanopore structures in materials. Adsorption and desorption isotherms of N_2 on the resorcinol-formaldehyde aerogels at 77 K are shown in **Figure 4**. Resorcinol-formaldehyde aerogels have type IV isotherms and a characteristic adsorption hysteresis, indicating that microporosity has had an important development regarding the mesoporosity.

The micropore and mesopore structural parameters can be determined from the N_2 adsorption isotherm, and in this case, the surface area of carbon resorcinol-formaldehyde aerogels was between 263 and 829 m²/g, mesopore volume was between 0.16 and 0.58 cm³/g, and the micropore volume was between 0.06 and 0.31 cm³/g (**Table 1**).

Sample		*D-A*				*B-J-H*	
	S_{BET} (m² g⁻¹)	V_m (cm³ g⁻¹)	E_0 (kJ mol⁻¹)	*n*	*L* (Å)	V_{meso} (cm³ g⁻¹)	*L* (Å)
Ae100	263	0.06	5.97	2.1	7.7	0.16	17.1
Ae400	455	0.23	5.35	3.4	8.1	0.10	17.1
Ae600	829	0.31	8.07	2.2	7.0	0.58	39.3

Table 1. Parameter values for the low pressure range (DA) and high pressures (BJH).

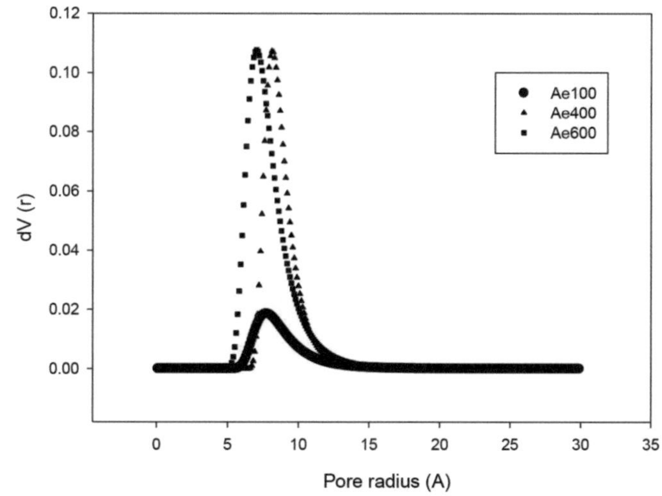

Figure 5. Pore size distribution D-A plot.

It means that the higher the ratio R/C increases, the Bet area increases, and hence the development of microporosity. According to the value obtained for the parameter *n* in the analysis using the D-A method for low pressures, a marked heterogeneity is evidenced by the porosity of the aerogels. **Figure 5** shows the pore distribution obtained for the samples.

Resorcinol-formaldehyde aerogels obtained have a homogeneous distribution of micropores and mesopores, which can be seen in the SEM micrograph. **Figure 6** shows that mesopores are relatively uniformly distributed in the resorcinol-formaldehyde aerogels.

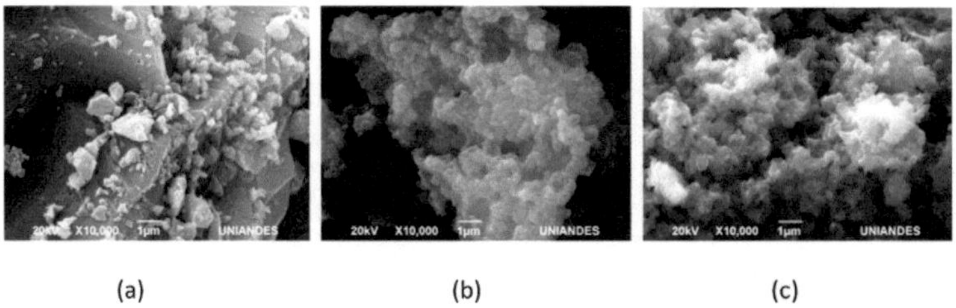

Figure 6. SEM image of resorcinol-formaldehyde aerogels: (a) R/C 100, (b) R/C 400, and (c) R/C 600.

The solution chemistry, pyrolysis temperature, and aerogel density affect the pore structures of carbon aerogels, and the ratio R/C also affects the chemical structure and the distribution of the groups in the surface of the aerogels.

A study to determine the effect of initial pH on the distribution of different groups onto the surface was conducted. This was performed by measuring the initial pH of synthesis as soon as all the reagents were mixed and polymerization was initiated. In addition, the pH point of zero charge (pH_{pzc}) was determined. Quantities of the aerogels weighed in a range from 0.050 to 0.300 g in about six samples in an interval of approximately 0.05 g, and each of the materials was placed in vials of 15 mL, and 10 mL of sodium chloride solution (0.1 N NaCl) was added to each of the materials. The vials were capped and left under constant stirring at a temperature of 298 K for 48 hours. The samples are centrifuged at 3000 rpm for 15 minutes. After that the pH of each sample was measured. The pH of each sample based on its weight is plotted, and the point of zero charge is determined as the pH at which the curve tends to all the samples in suspension. Surface chemical groups were followed according to the method of Boehm [15] (**Table 2**).

Sample	**pHi synthesis**	**Na_2CO_3 (lactones) (mEq/g)**	**$NaHCO_3$ (carboxylics) (mEq/g)**	**NaOH (phenol) (mEq/g)**	**$NaOC_2H_5$ (carbonyls) (mEq/g)**	**$HCl_{(basics)}$ (mEq/g)**	**pH_{pzc}**
Ae100	8.35	0.045	0.152	0.000	0.000	0.005	6.78
Ae400	7.16	0.041	0.217	0.000	0.000	0.000	6.75
Ae600	6.31	0.039	0.178	0.000	0.000	0.003	6.62

Table 2. Initial pH of synthesis with different surface groups and pHpzc carbon aerogels.

As can be seen, the initial pH of synthesis is slightly alkaline in character, which favors the formation of monomeric agglomerates (enolate ions) at the time of initiating polymerization.

Once aerogel resorcinol-formaldehyde is obtained, the pH_{pzc} is near neutrality, thereby inducing adsorption of cations. This variable has an effect on the adsorption capacity and allows explanation and optimization of adsorption, and this is consistent with the presence of surface groups consisting predominantly of lactones and carboxylic groups. This is very important when applying these materials to the adsorption of metal ions from aqueous solution.

The pollution of water resources due to the indiscriminate disposal of heavy metals has been causing worldwide concern for the last few decades. It is well known that some metals can have toxic or harmful effects on many forms of life. Metals, which are significantly toxic to human beings and ecological environments, include chromium (Cr), copper (Cu), lead (Pb), mercury (Hg), manganese (Mn), cadmium (Cd), nickel (Ni), zinc (Zn), and iron (Fe), etc.

Moreover, the removal of metal ions from their solutions in the presence of wastewaters may be due to the adsorption on surface and pores and also to complexation by these materials. The concentrations of some of the toxic metals such as nickel and chromium are higher than the permissible discharge levels in these effluents. Therefore, it becomes necessary to remove these heavy metals from these wastewaters by an appropriate treatment before releasing them into the environment.

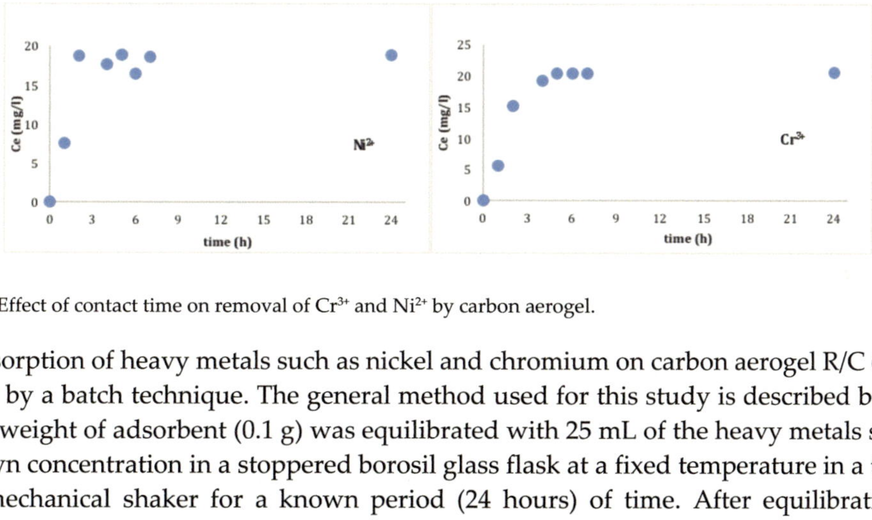

Figure 7. Effect of contact time on removal of Cr^{3+} and Ni^{2+} by carbon aerogel.

The adsorption of heavy metals such as nickel and chromium on carbon aerogel R/C 600 was studied by a batch technique. The general method used for this study is described below: A known weight of adsorbent (0.1 g) was equilibrated with 25 mL of the heavy metals solution of known concentration in a stoppered borosil glass flask at a fixed temperature in a thermostatic mechanical shaker for a known period (24 hours) of time. After equilibration, the suspension of the adsorbent was separated from the solution by filtration using Whatman No. 1 filter paper. The concentration of heavy metal ions remaining in the solution was measured by AAS using flame method. The pH of the adsorptive solutions was adjusted using nitric acid and sodium hydroxide at pH 5.0 for nickel and pH 2.0 for chromium.

It is observed that in two cases the percentage removal is comparatively lower for 3-hour contact time, with increasing removal efficiencies at higher contact time of 24 hours (**Figure 7**). It is evident from the results that the contact time required to attain equilibrium is dependent on the initial concentration of heavy metals. For the same concentration, the percentage removal of heavy metal increases with increase of contact time till equilibrium is attained. The optimal contact time to attain equilibrium with carbon aerogel was experimentally found to be about 6 hours.

Adsorption isotherm studies were carried out with different initial Ni^{2+} and Cr^{3+}concentrations (20–500 mg/L) and a fixed adsorbent concentration of 100 mg for 25.0 mL of solution. The equilibrium data obtained were analyzed in the light of Langmuir and Freundlich isotherms. For estimated residual metal concentration, C_e, the respective adsorbent metal uptake loading, q_e (mg of metal per g of carbon aerogel), for each sorption system was determined using the equation

$$q_e(\mathrm{mg/g}) = V(C_i - C_e)/W \quad (1)$$

where C_i is the initial metal concentration (mg/L) in a solution of volume V (L) and W is the mass of adsorbent (g).

Figure 8 shows the isotherm of nickel and chromium removal onto carbon aerogel. The following linearized forms of the Langmuir and Freundlich isotherm models were used in the present study to provide an objective framework to the generated equilibrium adsorption data. The Langmuir model allowing the calculation of maximum adsorption capacity, Q_o (mg/g), and the Langmuir constant, b (L/mg), is represented by the following equation:

$$C_e / q_e = 1/Q_o b + C_e / Q_o \quad (2)$$

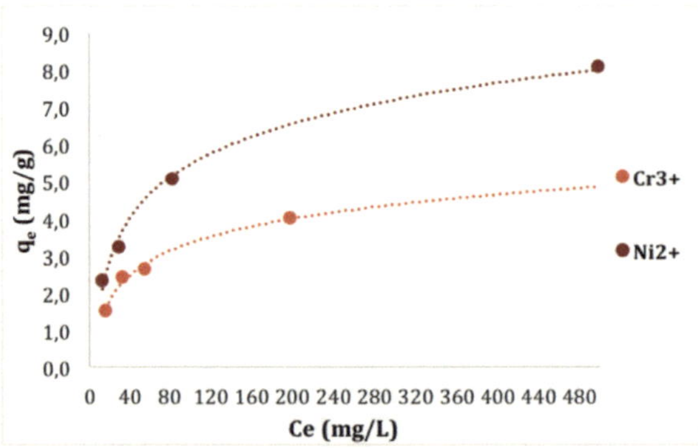

Figure 8. Isotherm curve for Cr^{3+} and Ni^{2+} adsorption onto carbon aerogel.

The linearized Freundlich model isotherm is represented by the following equation:

$$\log_{10}(q_e) = \log 10 K_f + 1/n(\log_{10} C_e) \quad (3)$$

where K_f and n are constants incorporating all the factors affecting the adsorption capacity and are an indication of the favorability of metal ion adsorption onto the adsorbent.

Adsorption capacity as indicated by value of "Q_o" is seen to be maximum for carbon aerogel, i.e., Ni^{2+} 7.602 mg/g with a much lower capacities for Cr^{3+}. The energies of adsorption, as indicated by "*b*" are seen to be highest for Ni^{2+} (**Table 3**). The values of 1/*n* lie between 1 and 10 indicating favorable adsorption.

	Freundlich			Langmuir		
	KE	1/*n*	R^2	Q_o (mg/g)	*b*	R^2
Ni^{2+}	11.55	1.265	0.8947	7.602	2.4125	0.9425
Cr^{3+}	10.52	1.209	0.9089	4.501	1.9874	0.9117

Table 3. Values of Langmuir and Freundlich isotherm constants for the adsorption of heavy metal ions on carbon aerogel.

2.2. Synthesis of Ti-SBA-15 by chemical vapor deposition (CVD) like a possible photocatalyst in degradation reactions

The physical and biological treatment are not sufficient for total pollutant degradation, for this reason the heterogeneous photocatalysis is been converted into an emergent technique. This technique is nonselective and can be used in the complex samples treatment. Moreover, the sun radiation is possible to be used as a primary energy source and the oxygen as an oxidant agent. This is a clear example of sustainable technology [16].

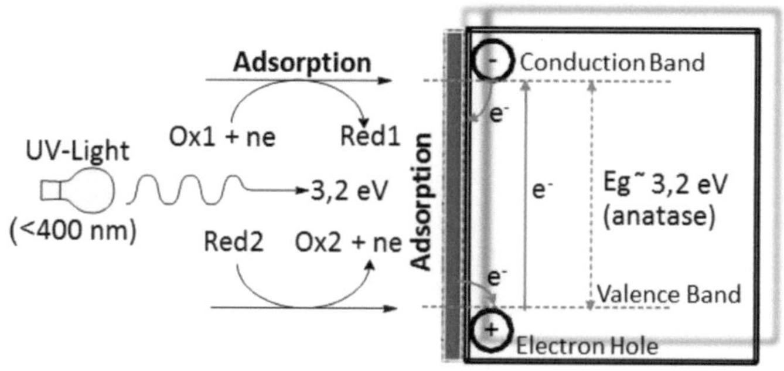

Figure 9. Energy band diagram TiO_2.

In the past decades, photocatalysis has been the research object in the contamination of water and air. The main advantages that were observed are as follows: work under environment conditions, complete destruction of pollutants and intermediaries, and low operating cost that confirmed their applicability in water treatment. However, there are many experimental and technical challenges that include (1) the commercial catalyst available presents low surface areas (factor determinant in the photocatalytic activity), (2) catalyst separation after treatment [17], (3) catalyst development with a strong absorption in the spectrum visible region [18], (4) understanding of the operation parameters of the photoreactor, and (5) mathematical inconsistencies in kinetic modeling.

In these processes, solid semiconductors are used, which are activated by UV light and TiO_2 (anatase), since it has a wide band-gap semiconductor (**Figure 9**) with different potential applications in catalysis, photovoltaics, optoelectronics, etc. [19, 20].

The TiO_2 is a good photocatalyst of organic compounds, where the absorbed light generates the formation of electron-hole pair (e^-/h^+). The h^+ interacts with the OH^- ions to generate hydroxyl radicals (OH•) that have a high oxidation potential and are capable of degrading organic compounds. Electrons are trapped by oxygen to produce superoxide radicals (O_2^-) [21].

Charge transfer and recombination processes occur simultaneously, and the electron-hole pair is recombined before reacting with adsorbed species on the catalyst surface [22]. The efficiency of the photocatalytic reaction depends on factors that determine the degree of oxidation of organic matter. Another determining factor is the nature of the contaminant, the pH of the medium, the wavelength of the radiation, the radiation flux, and the type and concentration of oxidant [23].

Nanoparticles (TiO_2) can be produced in different ways, for example, chemical vapor deposition, sol-gel, the oxidation of titanium tetrachloride, and titanium alkoxides hydrolysis [24]. In addition, it is very interesting to note that to improve its photocatalytic activity and obtain spherical particles with a high surface area, it supports mesoporous materials with porous structure.

Because the TiO_2 has a low specific surface area (10–30 m^2/g) and to increase the catalytic activity, it has been chosen to be used in mesoporous materials.

3. Mesoporous materials

SBA-15 (Santa Barbara-15) is considered an ordered mesoporous material with hexagonal channels of diameters between 3 and 14 nm, microporous walls (0.5–2 nm), and cylinder sizes (1–3 μm) (**Figure 10**).

Figure 10. Scanning electron microscope (SEM) SBA-15.

http://dx.doi.org/10.5772/64393

This mesoporous solids are synthesized using triblock copolymers like a poly(ethylene oxide)-poly(propylene oxide)-poly(ethylene oxide) (PEO-PPO-PEO) $EO_{20}PO_{70}EO_{20}$ Pluronic®-123 and tetraethyl orthosilicate (TEOS) as the silicate source.

The amphiphilic will form aggregates in aqueous environment due to the PEO chains as they are hydrophilic, while the PPO chains are hydrophobic. The processes that occur in aqueous solution are: (1) the formation of micelles due to the hydrophilic groups, which are oriented to the outer of the micelles and the hydrophobic is located in the interior, and (2) solubilization of block copolymers due to the interaction between water molecules and alkylene oxides through hydrogen bonds [24, 25].

In aqueous solutions, the silicate source (TEOS) is hydrolyzed when added with water and polymerizes to form a silica network (**Figures 11–13**) [26, 27].

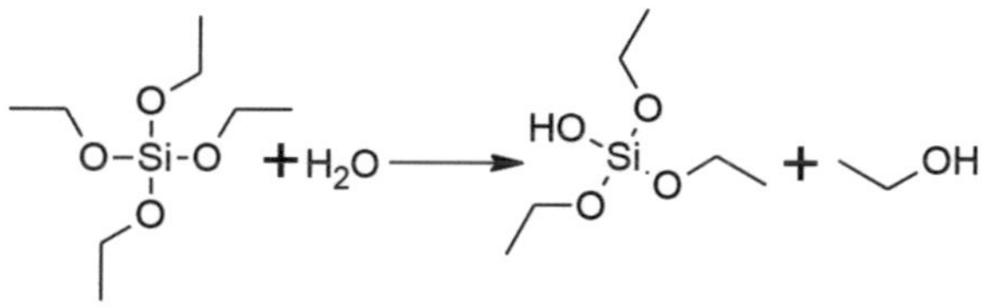

Figure 11. Hydrolysis of silica source.

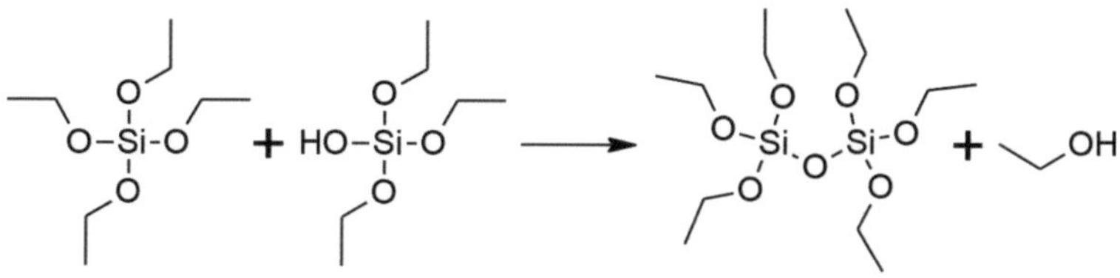

Figure 12. Silica network (1): alcohol condensation.

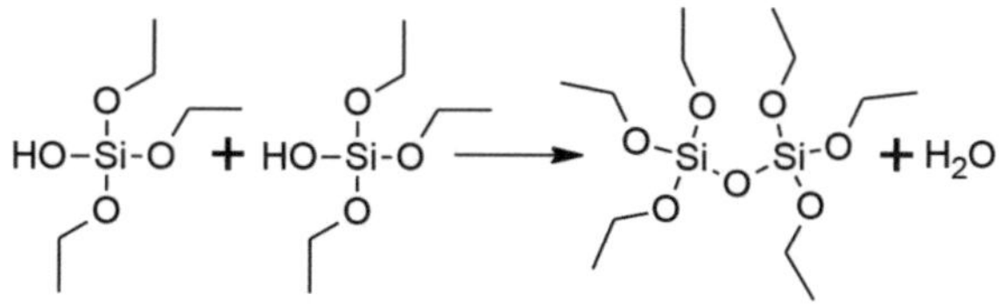

Figure 13. Silica network (2): water condensation.

The synthesis of SBA-15 is carried in acidic conditions because it is necessary to induce the interaction between triblock and TEOS. This can be controlled by pH in acid or basic conditions to modify the hydrolysis rate that is directly proportional to its concentration. In acidic medium, it increases the rate of hydrolysis, while in basic medium it increases the gelation rate. For this reason, the silica source is protonated (pH < 2, since pH = 2 is the isoelectronic point) because the interactions are important at the inorganic-organic interface. In acid media, the hydronium ions are associated with the alkylene oxygen atoms; in this way the long-range Coulombic interactions are added to the coassembly process. The cationic silica is a precursor, and the mechanism is carried out through an intermediate ionic like (S^0H^+) (X^-I^+) (**Figure 14**) [28]. The silicon atom can expand the coordination sphere and allow to coordinate with the anion (Cl^-).

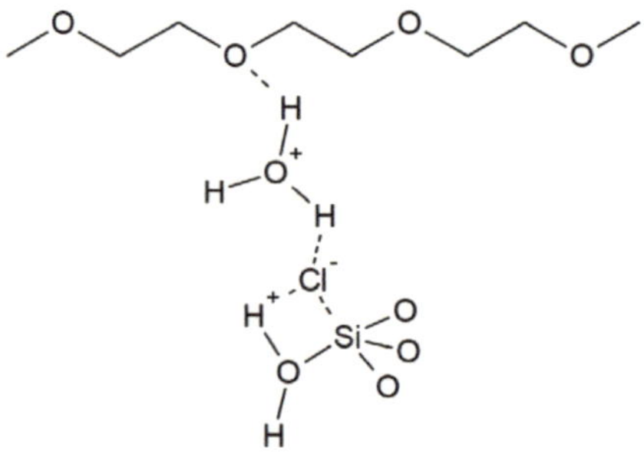

Figure 14. $S^0H^+X^-I^+$ interaction.

In SBA-15, the self-assembly route occurs through cooperative self-assembly, and the micellar rod cannot be formed before addition of the silica source. The micelles become elongated by the condensation of the silica source. The micellar rods arrange in a hexagonal pattern while the silica walls are built. The polymerization is simultaneous to the elongation and the silica precursor is polymerized on the PEO chains, and the water content in this area decreases to produce a change in the polarity resulting in a reduced curvature of the micelle (**Figure 15**) [29, 30].

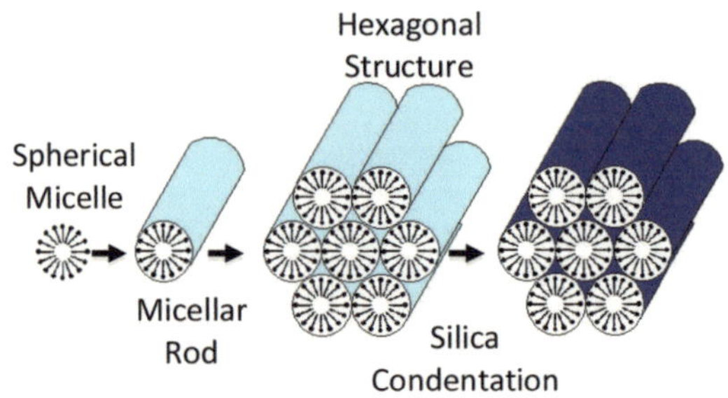

Figure 15. Schematic representation of SBA-15 before calcination.

The porous structures are not mesoporous (pore diameter > 2) because of the hexagonal arrangement of cylindrical aggregates. On the other hand, the PEO penetrates the pore walls, resulting in the microporous (pore diameter < 2 nm) production after calcination (**Figure 16**) [24, 25]. This process allows to remove the surfactants from SBA-15, generally the calcination temperature is 540°C for 6 hours with a rate of 5°C/min. The surfactant is decomposed between

100 and 300°C (**Figure 17**) and in this case the hexagonal structure retains its size. Above 300°C in air atmosphere, the polymers are combusted and the hexagonal framework is decreased due to its condensation. Approximately, in this stage it obtains a weight loss between 40% and 50%.

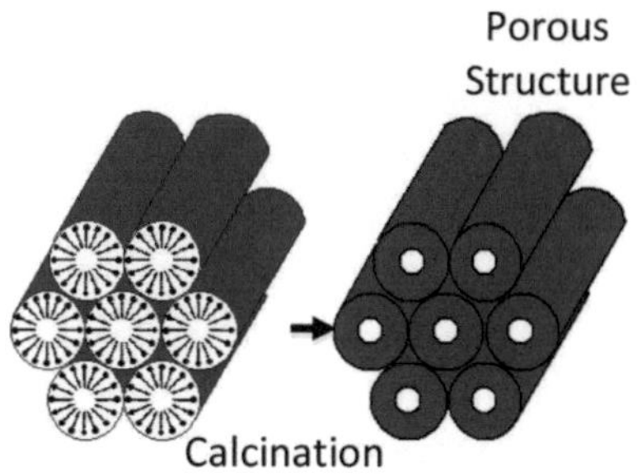

Figure 16. Schematic representation of SBA-15 after calcination.

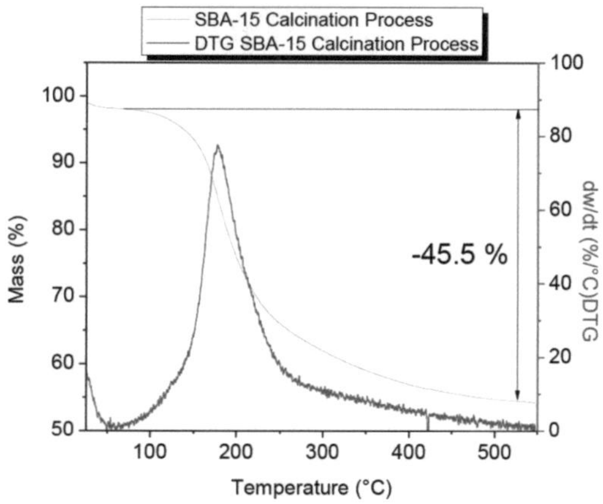

Figure 17. Thermogravimetric study in the elimination of surfactant.

These kinds of materials are activated with the inclusion of hetero-elements via postsynthesis or in situ methods that make them catalytically active [31, 32]. For this reason, the SBA-15 is considered one of the materials based on silica and it is considered promising because of its commercial availability, high surface area, thermal stability, and pore size distribution (PSD). On the other hand, it has been found that the addition of TiO_2 on these solids facilitates dispersion of anatase crystals, which are the essential active sites for degradation of contaminants and intermediaries. Ti-SBA-15 has been synthesized by chemical vapor deposition.

The CVD method for catalyst preparation has been defined as the process of deposition using reaction between surfaces sites such as OH groups and vapors of metal compounds. CVD is a useful method for the preparation of highly dispersed catalysts. The commercial Ti/SiO_2 catalyst was prepared by the impregnation of silica with $TiCl_4$ or an organic titanium compound followed by calcination. The active catalyst contains tetrahedral Ti^{IV} chemically bonded to siloxane ligands (≡SiO) [33].

In this way, the titanium is incorporated via postsynthesis. The experimental set up of CVD is depicted in (**Figure 18**). The SBA-15 is placed in a fixed-bed reactor and is first treated at 400°C in nitrogen flow, then brought back to the deposition temperature at 300°C. The nitrogen flow (150 mL/min) is saturated with $TiCl_4$ in a saturator. The saturator is immersed in a cooling bath at −10°C because of the vapor pressure of $TiCl_4$ at −10°C is 170 Pa (0.17%, V/V). The sample is then purged under nitrogen at 573 K, hydrolyzed at ambient temperature in moist air, dried at 80°C, and finally calcined in dry air at 673 K [34].

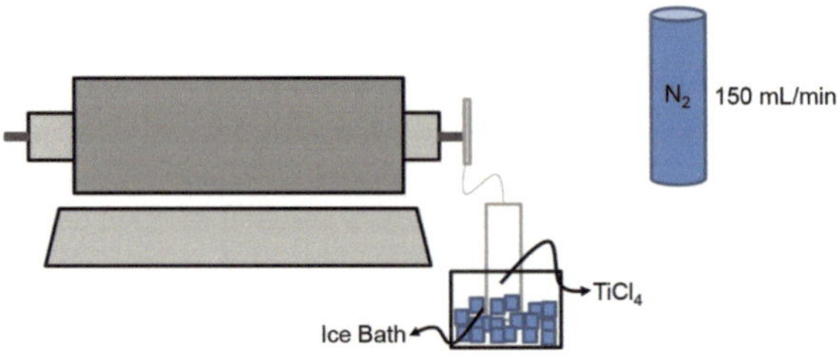

Figure 18. Experimental set up of chemical deposition vapor.

The reactions that can occur to the synthesized Ti-SBA-15 correspond to **Figure 19**.

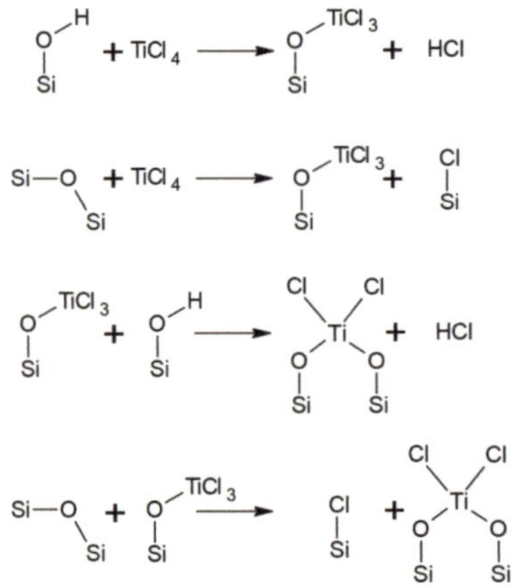

Figure 19. Chemical vapor deposition reaction $TiCl_4$-SBA-15.

The loss of the OH^- at 3750 cm^{-1} indicates that the free hydroxyl groups are reacting with the $TiCl_4$, when the deposition time is increased and the intensity of absorption peak is decreased (**Figure 20**). When a small amount of $TiCl_4$ is reacted with the silica, no change in the OH^- band intensity is observed. This is observed below 8 hours of reaction. The IR absorption peaks at 1230 and 1070 cm^{-1} are attributed to υ_{as} (Si-O-Si) vibrations of silica framework of SBA-15, and the peak at 802 cm^{-1} is attributed to υs (Si-O-Si) vibration [34]. The peak at 960 cm^{-1} belongs to v_{as} (Si-OH) or v_{as} (Si-OH) or vas (Si-O-Ti); this peak appears by two processes:

(1) The silanol groups are reacting more quickly than the hydroxyl groups.

(2) The $TiCl_4$ starts to react with the surface; there are many sites where the surface-attached $TiCl_3$ species can react further to give a bridged $TiCl_2$ species. This gives rise to an increase in the concentration of $TiCl_3$ on the surface. When the $TiCl_3$ does not have a neighboring hydroxyl groups the reaction is completed [33].

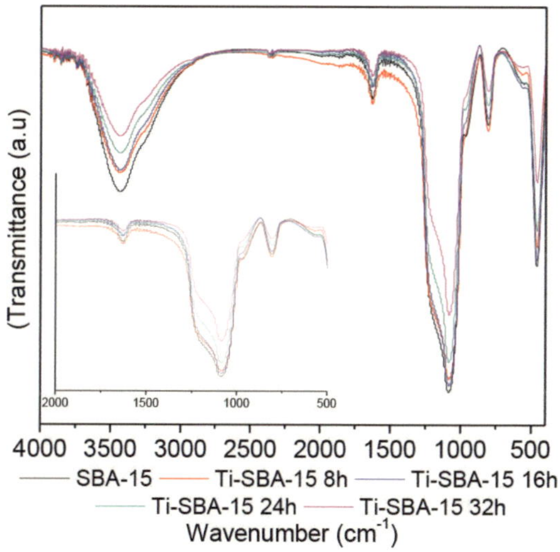

Figure 20. FT-IR spectra of Ti-SBA-15-h.

The relation between the peak at 960 and 802 cm^{-1} can be used to evaluate the amount of the Ti-O-Si bond. As shown in **Table 4**, when an increase in the deposition time is possible, deposit more Ti species on the SBA-15 surface.

	(Ti/Si) = z
Ti-SBA15 8h	1.7%
Ti-SBA15 16h	5.5%
Ti-SBA15 24h	9.4%
Ti-SBA15 32h	14.1%

Table 4. Intensity ratio Ti/Si.

Textural parameters of SBA-15 and Ti-SBA-15-h are calculated from N_2 adsorption isotherm at 77 K (**Figure 21**) and these are presented in **Table 1**. This type of porous solid exhibited type IV isotherm and the presence of a hysteresis loop [35], which are characteristics of mesoporous solids (2–50 nm pore size). Pore size distributions calculated with the Cylinder Pore NLDFT adsorption branch is reported in **Figure 22**. The PSD profile showed a broad peak with pore size >2 nm (mesoporous); this value corresponds at 3.4 nm to all samples.

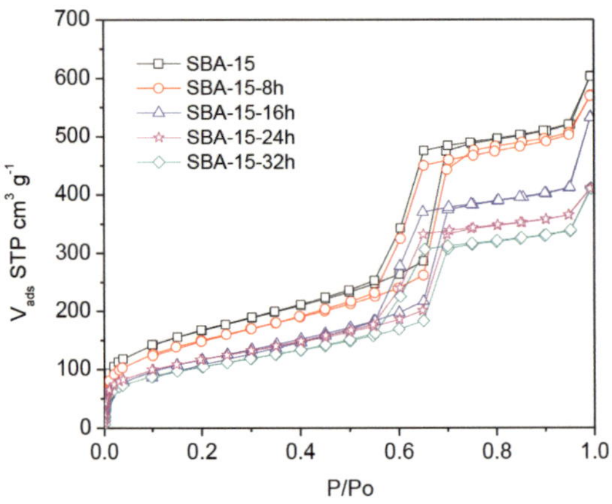

Figure 21. Isotherm adsorption N_2 77 K.

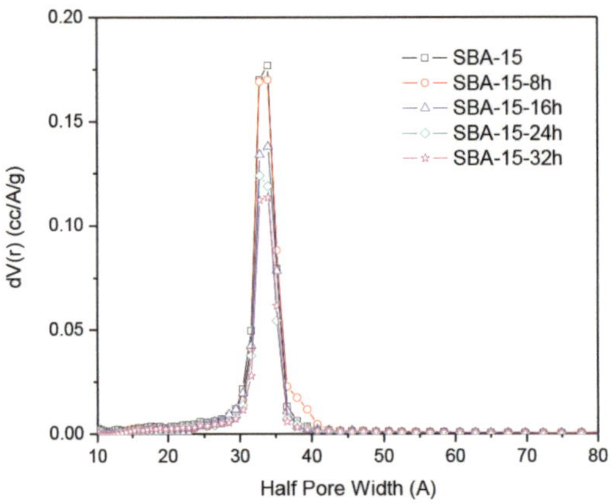

Figure 22. PSD obtained with N_2 77 K silica (cylinder pore, NLDFT adsorption branch).

	S_{BET} ($m^2 g^{-1}$)	$V_{micro\ DR}$ ($cm^3 g^{-1}$)	V_{meso} ($cm^3 g^{-1}$)	V_{Total} ($cm^3 g^{-1}$)	D_{MAX} (nm)
SBA-15	593	0.21	0.60	0.81	3.4
SBA-15-8h	537	0.19	0.59	0.78	3.4
SBA-15-16h	423	0.14	0.50	0.64	3.4
SBA-15-24h	414	0.15	0.41	0.56	3.3
SBA-15-32h	374	0.14	0.39	0.52	3.4

Table 5. Textural parameters N_2 adsorption data.

http://dx.doi.org/10.5772/64393

In **Table 5**, it is observed that the surface area is decreased when the deposition time is increased. The effects of CVD with $TiCl_4$ have a relation with the porosity, and the micropore volume decrease and is kept constant (0.14 cm^3/g) between 16 and 32 hours of deposition. The mesoporous volumes are equal for SBA-15 and Ti-SBA-15-8h, but when the deposition time is increased, the mesoporous volume decreases. This shows a similar behavior to the microporous volume. This behavior is obtained because the titanium species are supported on the surface of SBA-15. The shape of the isotherms obtained allows to deduce that there are no structural changes caused by chemical vapor deposition. Although the isotherms are parallel to the decrease in the total volume, it is due to the distribution of titanium species on the surface.

This relates to the results obtained in the IR, where Ti species on the surface increases, thereby decreasing the pore volumes. The effect of $TiCl_4$ on the surface has no effect on the textural properties.

The thermogravimetric stability was evaluated by observing that the inclusion of titanium does not affect this property because a strong interaction is developed between the titanium species and the surface (**Figure 23**). This is an important property for a photocatalyst that must be stable in a wide temperature range. The mass losses that originate are due to the calcination of the surfactant that was not removed during calcination and this allowed the porosity to be occluded (**Figure 24**).

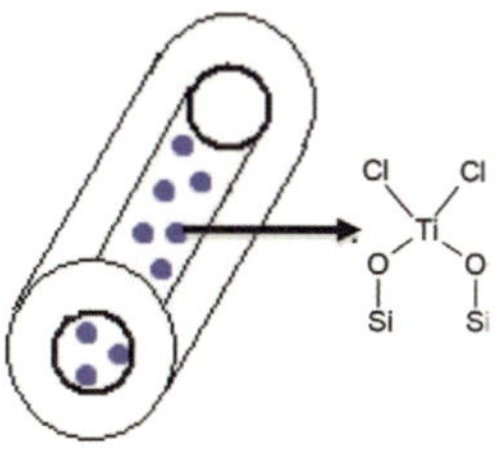

Figure 23. Schematic representation Ti-SBA-15-h.

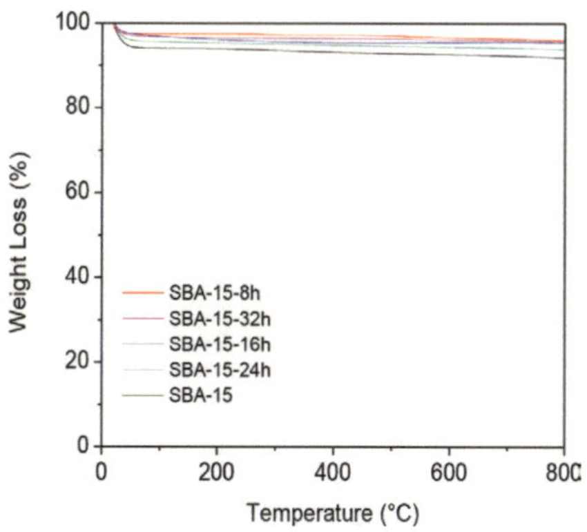

Figure 24. Thermogravimetric stability of Ti-SBA-15-h.

Scanning electron microscope images exhibit the morphology of SBA-15. The cylinders are the typical shape of this porous solid. The effect of chemical vapor deposition does not change the surface of the SBA morphology. The cylinders have a size of 1.4 μm and a width of 0.6 μm for all the samples (**Figure 25**). The chemical vapor deposition allows to obtain high dispersion titanium on the surface. In the SEM-EDX an increase in titanium dispersion can be observed when the deposition time is increased (**Figure 26**). These materials show a homogeneous distribution on the SBA-15 without apparent preferential concentration of titanium in some areas. The distribution is influenced for hexagonal framework of SBA-15.

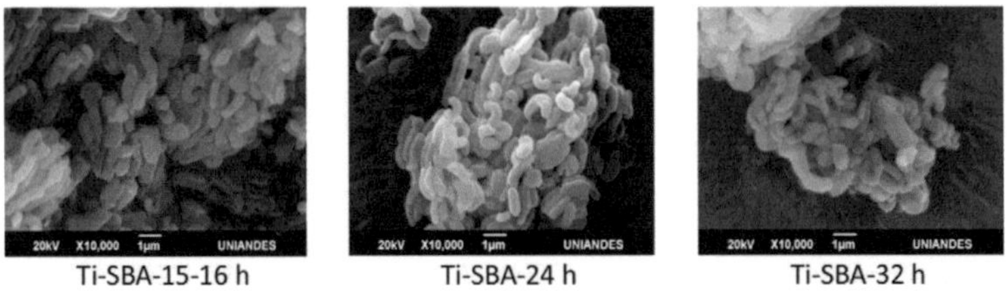

Figure 25. SEM of Ti-SBA-15-h at different deposition time.

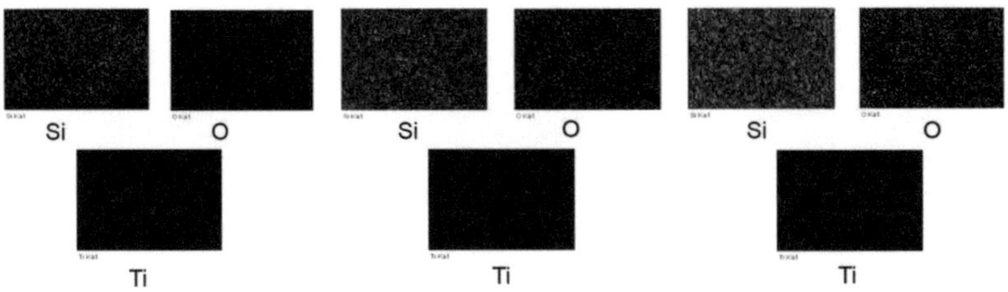

Figure 26. SEM-EDX dispersion titanium on the surface of SBA-15.

4. Conclusion

The development of new materials such as carbon aerogels for specific applications such as retention of metal ions from aqueous solution is a promising field of research, which has a great scope of further study. Developing materials with a specific texture and modifying the surface chemistry to increase the adsorption capacity are some issues for implementation in these materials.

The chemical vapor deposition is an excellent method to obtain Ti-SBA-15-h with a high disperse titanium on the SBA-15 surface. The TiO_2 is very important in photocatalytic process. Actually, different researches are putting efforts to obtain mesoporous materials with titanium to increase the yield in different reactions like photodegradation of organic compounds in water. This is an approximation to understand the synthesis of SBA-15, its mechanisms, and the way to include hetero-elements by chemical vapor deposition. It is necessary to further this

http://dx.doi.org/10.5772/64393

line of research with the aim to synthesize photocatalysts that are efficient, cheap, and reusable for the benefit of the environment.

Acknowledgements

The authors thank the University of the Andes (Bogota, Colombia) and National University of Colombia (Bogotá) by the framework agreement between the two institutions under which this research was developed. Special thanks to the Faculty of Science and the Vice-Rectory of Research at the University of the Andes (Bogota, Colombia) for funding this research and the "Semilla Project" 2016-1 of Faculty of Sciences, University of the Andes (Bogota, Colombia).

Author details

Rafael Alberto Fonseca-Correa[1], Yesid Sneider Murillo-Acevedo[1], Liliana Giraldo-Gutiérrez[2] and Juan Carlos Moreno-Piraján[1*]

*Address all correspondence to: jumoreno@uniandes.edu.co

1 Department of Chemistry, Research Group Porous Solids and Calorimetry, Faculty of Sciences, University of the Andes, Bogotá, Colombia

2 Department of Chemistry, Faculty of Sciences, National University of Colombia, Bogotá, Colombia

References

[1] Zulamita Z.B., Carrasco M.F., Vicente J., Moreno C.C. Langmuir. 2013; 29: 6166–6173.

[2] Yong K., Ya Z., Xiaodeng S., Sheng C., Mong Y., Kaiming T., Junjun Z. Journal of Non-Crystalline Solids. 2012; 358: 3150–3155.

[3] Geng P.W., Junbing Y., Dapeng W., Rui X., Khalil A., Chung X.L. Materials Letters. 2014; 115: 1–4.

[4] Jie Z., Rang L., Weiling N., Yanjun W., Xinglung G. International Journal of Hydrogen Energy. 2013; 38: 10864–10870.

[5] Zapata B.Z., Moreno C.C., Carrasco M.F. Journal of Sol-Gel Science and Technology. 2015; 75 (2): 407–412.

[6] Pierre A.C., Pajonk G.M. Chemical Reviews. 2002; 102: 4243–4265.

[7] Pekala, R.W. US4873218 (1989).

[8] Pekala, R.W. US4997804 (1991).

[9] Niaping L., Jun S., Dong L. Microporous and Mesoporous Materials. 2013; 167: 176–181.

[10] Moreno C.C., Maldonado H.F.J. Carbon. 2005; 43: 455–465.

[11] Brinker C.J., Scherer G.W. Sol-Gel Science: The Physics and Chemistry of Sol-Gel Processing. Academic Press: San Diego, CA, USA, 1990.

[12] Ai D., Bin Z., Zhihua Z., Jun S. Materials. 2013; 6: 941–968.

[13] Yousheng T., Morinobu E., Katsumi K. Recent Patents on Chemical Engineering. 2008; 1 (3): 192–200.

[14] Alain C.P., Gérard M.P. Chemical Reviews. 2002; 102 (11): 4243–4265.

[15] Boehm H.P. Carbon. 2002; 40: 145–149.

[16] Laoufi N.A., Tassalit D., Bentahar F. Global Nest Journal. 2008; 10: 404–418.

[17] Kumar R.P., Venkata L.R.P., Kwon E., Ki H.K., Tahmina A., Sudhakar K. Environment International. 2016; 91: 94–103.

[18] Hou Y., Qiujun L., Wang H., Haitao L., Youyu Z., Shiying Z. Materials Letters. 2016; 173: 13–17.

[19] Linsebigler A., Lu G., Yates J.T. Chemical Reviews. 1995; 95: 735–758.

[20] Fujishima A., Rao N.T., Rryk D.A. Journal of Photochemistry and Photobiology C: Photochemistry Reviews. 2000; 1: 1.

[21] Lin X H., Wu Y., Xiang J., He D., Li S.F.Y, App. Catal B: Environ. 2016; 74, 64–74.

[22] Prieto R.L., Oller I., Klamerth N., Agüera A., Rodríguez E.M., Malato S. Water Research. 2013; 47: 1521–1528.

[23] Nan C.M., Jin B., Chow C.W.K., Saint C. Water Research. 2010; 44: 2997–3027.

[24] Impéror C.M., Davidson P., Davidson A. Journal of the American Chemical Society. 2000; 122:11925–11933.

[25] Flodstrom K., Alfredson V. Microporous and Mesoporous Materials. 2003; 59: 167–176.

[26] Brinker C.J. Journal of Non-Crystalline Solids. 1988; 100: 31–50.

[27] Hench L.L., West J.K. Chemical Reviews. 1990; 90: 33–72.

[28] Meynen V., Cool P., Vasant E.F. Microporous and Mesoporous Materials. 2007; 104: 26–38.

[29] Corma A. Chemical Reviews. 1997; 97: 2373–2419.

[30] Kresge C.T., Leonowicz M.E., Roth W.J., Vartuli J.C., Beck, J.S. Nature. 1992; 359: 710–712.

[31] Segura Y., Meynen V., Van der Voort P., Mees F., Vasant E.F. Journal of Physical Chemistry B. 2004; 108: 3794–3800.

[32] Chen Y.Y., Huang Y.L., Xiu J.H., Han X.W., Bao X.H. Applied Catalysis A: General. 2004; 273: 185–191.

[33] Kinney J., Staley R. Journal of Physical Chemistry. 1983; 87: 3735–3740.

[34] Junming Y., Wangcheng Z., Xiaohui L., Yanglong G., Yanqin W., Yun G., Guanzhong L. Microporous and Mesoporous Materials. 2012; 148: 131–136.

[35] Oguraa M., Guillet N.R., Brouri D., Casale S., Blanchard J., Cychosz K., Thommes M., Thomas C. Microporous and Mesoporous Materials. 2016; 225: 440–449.